SCIENCE

HACKS

100 clever ways to help you understand and remember the most important theories

COLIN BARRAS

An Hachette UK Company
www.hachette.co.uk

First published in Great Britain in 2018 by Cassell, a division of
Octopus Publishing Group Ltd
Carmelite House
50 Victoria Embankment
London EC4Y 0DZ
www.octopusbooks.co.uk
www.octopusbooksusa.com

Distributed in the US by
Hachette Book Group
1290 Avenue of the Americas
4th and 5th Floors
New York, NY 10104

Distributed in Canada by
Canadian Manda Group
664 Annette St.
Toronto, Ontario, Canada M6S 2C8

ISBN 978-1-84403-984-5

A CIP catalogue record for this book is available from the British Library.

Printed and bound in China

10 9 8 7 6 5 4 3 2 1

Publishing Director: Trevor Davies
Senior Editor: Alex Stetter
Junior Designer: Jack Storey
Design and layout: Simon Buchanan, Design 23
Illustrators: Design 23
Production Controller: Sarah Kulasek-Boyd

Contents

Introduction

For some students in North America, the first geology lesson is about poker. For others, it focuses on pearls. My British geology teacher opted to talk about camels. In all three cases, the idea is to make sure the students learn the main geological divisions of the last half a billion years or so – Cambrian, Ordovician, Silurian, and so on. The details vary, but the method of choice is the same: a mnemonic.

Come over some day, maybe play poker. Three jacks can take queens.

Cold oysters seldom develop many precious pearls, their juices congeal too quickly.

In the UK, where the geological divisions chosen are slightly different, the mnemonics run something like this: *Camels often sit down carefully. Perhaps their joints creak? Early oiling might prevent permanent rheumatism.*

Science is packed full of these phrases. There are mnemonics for remembering the order of the planets of the Solar System, the chemical elements of the periodic table and the different taxonomic ranks used to classify organisms. But these are not the only little hacks scientists and science students can use to memorize key bits of information. One of the most effective ways to remember a scientific theory or law is to boil it down to a pithy phrase.

Within a few years of the publication in 1859 of *On the Origin of Species*, Charles Darwin's theory of evolution by natural selection had been reduced to a simple expression: survival of the fittest.

Another of the great mid-19th century scientific achievements was the formulation of the first and second laws of thermodynamics. Today, many physicists and non-physicists alike sum them both up in one memorable witticism: you cannot win, and you cannot break even.

Reduce a core scientific concept too far, though, and you can run into trouble. Some people dismiss "survival of the fittest" as a tautology (they argue it could be rewritten as "survival of those that survive"). The single sentence summary of the first and second

laws of thermodynamics, meanwhile, is rather cryptic to anyone who does not already know what the two laws are. In this book, I have aimed for a middle ground. The hacks that follow may not be as clever or instantly memorable as the famous examples above, but I hope they are a little more useful – especially for those who want to find a shortcut to understanding some of the most challenging concepts in science.

No.1

The theory of evolution by natural selection

Why Darwin matters

Charles Darwin // 1809–1882

1/Helicopter view: Most people have heard of **Charles Darwin**. In the 1830s, Darwin traveled to exotic lands on the Royal Navy ship HMS *Beagle*. Among many observations he made, Darwin noticed how much variability there is in the natural world, and the difficult struggle for existence individual organisms face.

A couple of years after his return to England, Darwin read an essay by economist **Thomas Malthus**, which painted a bleak picture of humanity's future: populations could double with every passing generation, but food production would fail to keep pace, leading to starvation for many – another struggle to survive.

Darwin suspected that a Malthusian-like struggle plays out in the natural world, and that this struggle could form the basis of a mechanism through which new species evolve.

Perhaps surprisingly, Darwin didn't rush to publish. He shared his ideas with friends, who suggested the argument could be made stronger with a larger body of supporting evidence. Darwin took the advice, and spent several years studying species to gather such evidence.

Then, in the 1850s, he received word from another scientist working in Indonesia. Independently, **Alfred Russel Wallace** was homing in on a very similar view of the evolutionary process. In 1858, a scientific society in London heard letters from both Darwin and Wallace explaining the new idea. The following year, Darwin published a lengthy book setting out his evidence for **evolution by natural selection**. *On the Origin of Species* became famous. Darwin's reputation was sealed.

Darwin was influenced by his voyage on the *Beagle* and reading Malthus's work.

2/ Shortcut: Darwin realized that organisms often give birth to large numbers of offspring. There is variability among those offspring: some might have slightly longer limbs or sharper eyes, for instance. Most individuals will struggle to survive, but a lucky few will have features that make it easier for them to flourish, so they will prosper and breed – they will be naturally selected. Darwin assumed these individuals' offspring would inherit some of the beneficial features, and so over time these features would become more common in the population. Gradually, the population will evolve into a new species characterized by the new features.

See also //

2 The principle of coevolution, p.8

5 The modern synthesis, p.14

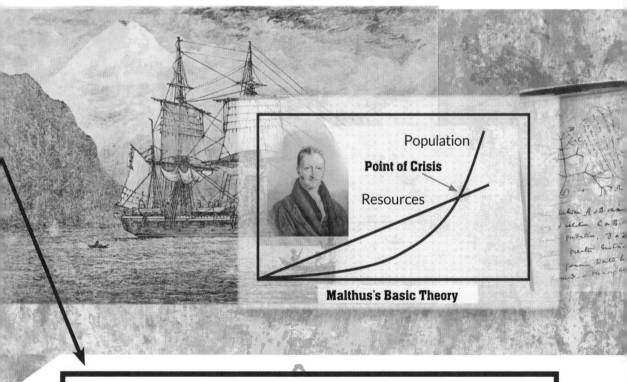

Population

Point of Crisis

Resources

Malthus's Basic Theory

3/ Hack: Some individuals are just inherently better suited to their environment than others. They are more likely to survive and breed.

Consequently, they have more influence on the evolutionary trajectory of their lineage.

No.2

The principle of coevolution
Darwin's astonishing predictive powers

Gaston de Saporta // 1823–1895

1/ Helicopter view: In the years following the publication of *On the Origin of Species*, **Charles Darwin** became famous for his work on evolution. But he also had a reputation for his specialist knowledge – including on the biology of orchids.

In 1862 Darwin was sent a specimen of an unusual Madagascan orchid, which produced flowers with nectar at the base of a 25-cm- (10-in-) long trumpet. Darwin made a prediction: Madagascar must be home to a species of insect with a tongue that was 25cm (10in) long. His prediction was based on the idea that two or more species influence each other's evolution, or coevolve – although the term "coevolution" wasn't coined until the 1960s. Darwin might never have used the word, but he was conscious of the importance that the **principle of coevolution** might play in nature.

A famous example involves the initial appearance of flowering plants. In Darwin's time the fossil record suggested that flowers burst onto the scene in the geological equivalent of the blink of an eye. This bothered Darwin, who believed that species evolved very gradually. **Gaston de Saporta** suggested the mystery might be explained by what scientists would now call coevolution. Perhaps flowers evolved so rapidly because they coevolved with insect pollinators, and this sped up the usually slow process of evolution. Darwin thought de Saporta's idea was "splendid" (although scientists now think it was probably wrong, not least because fossils found since Darwin's day show flowers evolved more gradually than once thought).

And what of Darwin's prediction? About 20 years after his death, biologists found a species of Madagascan moth with an extraordinarily long tongue. In 1992, scientists confirmed that the moth does indeed feed on nectar from the unusual orchid.

Morgan's sphinx moth uses its astonishingly long proboscis to feed on Madagascar's remarkable orchids.

2/Shortcut: Darwin suggested that some species of flowering plant have entered into an "evolutionary pact" with some species of insect. Flowers produce sweet nectar that provides insects with nourishment, while the insects (unwittingly) carry pollen between flowers as they sup on the nectar, helping the plants fertilize each other and produce viable seeds. If a plant has evolved flowers with difficult-to-reach nectar it stands to reason that an insect must have evolved a tongue to take advantage of the food.

See also //

8 The Red Queen's hypothesis, p.20

3/Hack: Biological species don't evolve in a vacuum. The evolutionary path a species takes is modified by its environment.

This means two or more species in the environment might end up influencing each other's evolution.

No.3

The Lamarckian theory of inheritance

A (not entirely) wrong way to view evolution

Jean-Baptiste Lamarck // 1744–1829

1/ Helicopter view: **Charles Darwin** was not the first scientist to think about evolution. Of those who came before him, one is particularly famous: **Jean-Baptiste Lamarck**. In the early 19th century, Lamarck came up with a complicated and detailed evolutionary framework in which species were compelled to gain complexity over time. However, there are two concepts he championed as part of this framework for which he is now famous. The first is the idea that organisms change during life in response to their environment – they *acquire* new characteristics – the second is that the organism's offspring *inherit* these acquired changes. Decades after Lamarck's death, these two ideas became known as the **Lamarckian theory of inheritance**.

Perhaps surprisingly, Lamarckism remained popular even after Darwin had published *On the Origin of Species* in 1859. In fact, in the late 19th century and early 20th century many scientists were sceptical about Darwin's **theory of evolution by natural selection**. During this period – sometimes called the "Eclipse of Darwin" – many argued that evolution was actually driven by the inheritance of acquired characteristics. In essence, they argued that evolution was Lamarckian not Darwinian. It was only when scientists rediscovered the work of **Gregor Mendel** and began to see evolution through the prism of genetics that Darwin's idea of natural selection became popular again (see: **The modern synthesis**, page 14).

There is a postscript to the story. Genetic studies largely disproved Lamarckism. In the last few years, though, strange exceptions to this rule have emerged. For instance, some bacteria "hardwire" life experiences into their DNA, meaning their offspring do in fact inherit features acquired by the parent cell. In some circumstances Lamarckism apparently does occur.

The giraffe's long neck – an evolutionary puzzle Lamarck's theory seemed to solve.

2/ Shortcut: One way to think about the broadly disproved Lamarckian theory involves giraffes. During its life, the (short-necked) ancestor of the giraffe strained to nibble leaves from tall trees. This literally stretched the animal's neck, making it slightly longer. Lamarckism suggests that the giraffe's offspring inherited this feature: at birth, their necks were slightly longer than their parents had been when they were born. Later in life, these offspring also strained to feed, so their necks stretched a little more. When they reproduced, their offspring began life with even longer necks. Countless generations of this process gave rise to the modern giraffe, with a neck that can be almost 2 metres (6 feet 6 inches) long.

See also //

1 Theory of evolution by natural selection, p.6

16 The theory of Mendelian inheritance, p.36

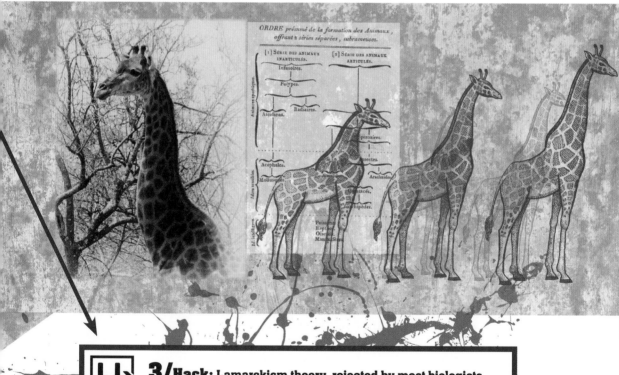

3/ Hack: Lamarckism theory, rejected by most biologists, suggests that organisms adapt to their environment throughout life and then pass these acquired changes on to their offspring.

It suggests evolution is driven by experiences gained in life.

No.4

The concept of neoteny

Why humans are forever young

Julius Kollmann // 1834–1918

 I/Helicopter view: In the 1860s **Auguste Duméril** made an astonishing discovery. He had received a shipment of strange amphibians from Mexico. Scientists knew that the animal – the axolotl – grew into a peculiar aquatic species: what Duméril discovered was that if the animal was forced to live in drier environments it metamorphosed into something much more like a salamander, with lungs instead of gills. Duméril's work was a step towards recognizing an important factor at play in evolution.

Almost at the same time that Duméril was performing his experiments, **Edward Drinker Cope** was considering the role that accelerated or arrested development might play in the origin of new species. Organisms can change their appearance drastically as they grow to sexual maturity. Perhaps, suggested Cope, some new species might evolve simply by random shifts in this pace of development. **Charles Darwin** read Cope's essay on this concept and was clearly impressed. He folded the idea into his sixth and final edition of *On the Origin of Species*, published in 1872.

A decade later **Julius Kollmann** recognized that Duméril's work on the axolotl provided a concrete example of Cope's concept in action. Duméril had helped to show that the axolotl seems to become stuck in a state of arrested development if it spends its entire life in water: it retains gills and other juvenile features even as a sexually mature adult. Kollmann called this retention of juvenile characters into adulthood the **concept of neoteny**. Today, many biologists recognize that neoteny has played a key role in the evolution of species – probably including our own.

Auguste Duméril // 1812–1870

Raise the axolotl in water and it will retain the features of a juvenile amphibian even into old age.

2/Shortcut: Biologists often use humans as a classic example of neoteny. Our nearest living relatives, the chimpanzees and bonobos, produce babies that look relatively similar to human babies. However, as chimps and bonobos grow they develop additional features: they gain more hair, a large jaw and long muscular arms, for instance. As humans grow they retain many "babyish" features: little hair, a small and weak jaw, relatively short arms, and so on. Humans certainly grow old, but in a peculiar way we look forever young.

See also //

1 The theory of evolution by natural selection, p.6

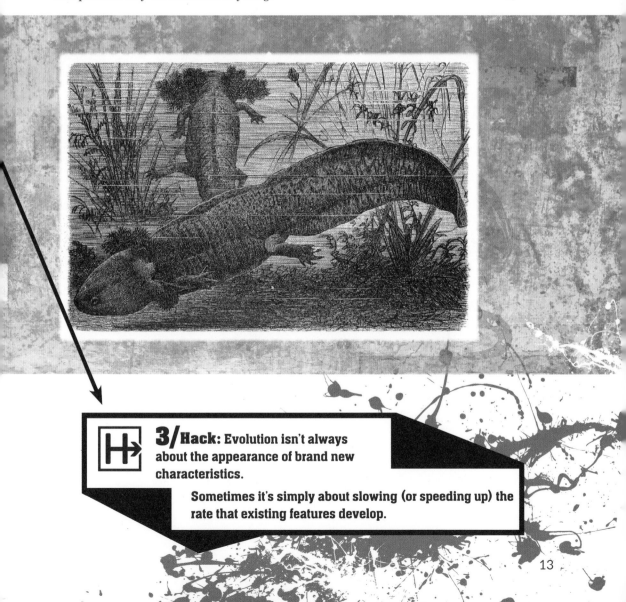

3/Hack: Evolution isn't always about the appearance of brand new characteristics.

Sometimes it's simply about slowing (or speeding up) the rate that existing features develop.

No.5
The modern synthesis
Evolution as we now know it

J B S Haldane //
1892–1964

1/ Helicopter view: Charles Darwin first outlined his **theory of evolution by natural selection** in the 1850s – but 40 years later, biologists were still arguing about its merits. At the turn of the 20th century biologists began to recognize the important role that genetics play in evolution, eventually realizing that genes supported Darwin's views on evolution and leading to what is called the **modern synthesis**.

Just a few years after Darwin published *On the Origin of Species*, **Gregor Mendel** began experimenting on pea plants and exploring the way organisms inherit features from their parents. Mendel was ahead of his time – it wasn't until the 20th century that biologists took his ideas seriously and the study of genetics began in earnest.

But it was vital that they did. Genetics provided a missing piece in Darwin's theory – a mechanism through which characteristics could be inherited. Even so, many early geneticists thought their work actually disproved Darwin's ideas. Genes were discrete chunks of heritable information, and they seemed to imply evolution occurred in abrupt leaps: if a gene for "tallness" emerged, for instance, then a population might evolve rapidly into a new "tall" species. Darwin had insisted that evolution was slow and steady, with new species evolving gradually (see: **The theory of punctuated equilibrium**, page 22).

But as geneticists began to probe more deeply they changed their minds. In the 1920s they discovered that a feature like "tallness" is often coded by dozens of genes, not just one. Scientists like **Ronald Fisher**, **J B S Haldane** and **Sewall Wright** began to produce mathematical models to explore how populations might evolve if one or more individuals happened to be born with a beneficial new genetic mutation. They realized that genetic complexity led to a pattern and pace of evolution that was perfectly compatible with Darwin's concept of natural selection. More than 60 years after the publication of *On the Origin of Species*, biologists had finally accepted Darwin's ideas.

Digital images can look smooth at a distance even though they look blocky in detail.

2/ Shortcut: One of the key revelations of the modern synthesis was that biological features are often controlled by dozens of genes, not just one. This simple discovery helped explain how discrete, binary genes can build organisms that seem to seamlessly blend features from both parents and evolve gradually into new forms. A computer game character looks blocky when it is drawn on a 12 by 16 pixel grid but super-smooth if it is drawn using thousands of pixels: similarly, an organism — and evolution — has a "smooth" appearance because organisms are often built from thousands of genes, not a mere handful.

See also //

1 The theory of evolution by natural selection, p.6

16 The theory of Mendelian inheritance, p.36

3/ Hack: The modern synthesis showed how evolution by natural selection works through genetic inheritance.

It represents the moment that biologists agreed about what evolution is and how it occurs.

No.6
Sexy son hypothesis

Why females find cheating males attractive

Ronald Fisher // 1890-1962

1/ Helicopter view: In the 19th century, **Charles Darwin** and **Alfred Russel Wallace** were both fascinated by the striking appearance of some animals, particularly males. The **theory of natural selection** did not seem to be able to explain why male peacocks have such elaborate tails or why stags have large antlers. Something else must be going on.

By the 1870s, Darwin had come up with a working hypothesis: such traits were a consequence of a special type of natural selection called sexual selection. Natural selection is based on the idea that a species evolves to cope with pressures in its environment – for instance, some animals will evolve leaner, more muscular bodies to outpace a natural predator. Sexual selection suggests pressures can come from within the species too. If, for example, females of the species prefer to mate with males that have brighter fur or feathers, then this preference might eventually lead to the evolution of brighter males.

An important point is that the pressure to be sexually desirable can be so strong that males might evolve features that appear to reduce their chances of survival. It seems strange for a male bird to evolve red feathers that are conspicuous to predators, but the pros of sexual attractiveness and mating opportunities might outweigh the cons of being more likely to be eaten.

Ronald Fisher refined the idea in 1930. By this point, the concepts of natural selection and genetic inheritance had come together into a powerful evolutionary model. This meant that biologists had begun to think of organisms and evolution in terms of genes. Fisher realized that females might boost the chances of their genes surviving and multiplying by choosing to mate with attractive males, precisely because any male offspring from these unions may inherit their father's attractiveness and his likelihood of finding many opportunities to mate. This updated view of sexual selection became known as the **sexy son hypothesis**.

The male peacock may have evolved its elaborate tail simply to attract females.

2/Shortcut: Fisher's sexy son hypothesis makes sense of one of life's enduring mysteries: why females of the species (including the human species) often seem to be attracted to unfaithful males. Females of many species seem to accept that they must devote time to raising their infants. This means they will have relatively few children and their genes won't spread far. But if male children inherit their father's attractiveness and his willingness to sleep around, the female should end up having dozens of grandchildren. Ultimately, her genes (many of which are carried by her sexy sons) will multiply.

See also //

7 The concept of kin selection, p.18

28 The grandmother hypothesis, p.60

3/Hack: Females mate with attractive but promiscuous males in the expectation that the sons from such unions will inherit their father's attractiveness and his licentiousness.

These "sexy sons" should spread their genes (which include many of their mother's genes) far and wide.

The concept of kin selection

Why individuals matter less than you might think

John Maynard Smith// 1920–2004

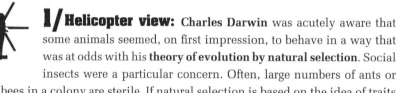

1/ Helicopter view: **Charles Darwin** was acutely aware that some animals seemed, on first impression, to behave in a way that was at odds with his **theory of evolution by natural selection**. Social insects were a particular concern. Often, large numbers of ants or bees in a colony are sterile. If natural selection is based on the idea of traits passed down the generations, how could the trait of "sterility" survive and prosper? In his book *On the Origin of Species*, Darwin suggested an explanation.

Darwin began by speculating that some fertile animals may carry the means to produce sterile offspring if they need to. He then went on to suggest that such sterile offspring might, through their activities, help their fertile peers to prosper. If they did so, then the animal population as a whole could thrive and evolve. In a funny, indirect way, sterile worker ants would help perpetuate the survival of sterile worker ants – using fertile ants as a conduit.

In the 1930s, once biologists had married the science of genetics and the science of natural selection (see: **The modern synthesis**, page 14), they modified this idea slightly. **Ronald Fisher** and **JBS Haldane** reasoned that evolution is really about the survival of genes rather than individuals (see: **Selfish gene theory**, page 46). The insects in a colony are closely related, which means sterile individuals carry very similar genes to their fertile siblings. If sterile workers altruistically help fertile individuals to survive and breed, the sterile workers "win" too – genes very similar to their own survive and replicate through their fertile peers. The concept grew in popularity. In the 1960s, **John Maynard Smith** gave it a name: **kin selection**.

Kin selection provides a solution to the mystery of why some ants are born sterile.

2/Shortcut: To understand the concept of kin selection it might help to look at the human body from a different perspective. Imagine one of your toes becomes infected with a nasty pathogen. The doctors give you a choice: have the toe amputated or let the infection grow and spread, putting your life at risk. Clearly, amputation is the best option, even though this involves "sacrificing" all of the human cells in the toe. Now imagine each of those cells is actually an individual animal living in a tightly knit family group. It makes evolutionary sense for some of those animals to sacrifice themselves if doing so ensures the survival of the group as a whole.

See also //

1 Theory of evolution by natural selection, p.6

3/Hack: It's what's on the inside that counts. If two or more individuals have a large number of genes in common then, from an evolutionary perspective, these multiple individuals are effectively a single entity – a kin group.

No.8

The Red Queen's Hypothesis
Evolving to stand still

Leigh Van Valen// 1935–2010

 1/Helicopter view: Evolution is about survival. As such, we might expect species would get "better" at surviving as time passes. Or to put it another way, species should have become better at avoiding extinction. It's an idea that **Leigh Van Valen** explored in the 1970s, and he discovered something unexpected.

When he looked at the fossil record and discounted dramatic extinction events, Van Valen discovered that extinction rates have remained more or less the same in any given group of animals. This means, for instance, that the probability of a mammal species going extinct is roughly the same today as it was 40 million years ago. Mammal species have apparently got no better at surviving, even after millions of years of evolution. Van Valen looked for an explanation. The one he came up with is known as the **Red Queen's hypothesis**.

Van Valen's hypothesis leaned on the idea that species don't live in a vacuum. Two or more species in an ecosystem interact, and influence each other's evolution. **Coevolution** is often an antagonistic process: one species does well at the expense of another, and vice versa. This means that if one species gains a survival advantage by evolving a new trait, another species probably finds its chances of survival disadvantaged as a consequence. In other words, the survival odds of one species are constantly being undermined by evolutionary innovations that occur in other species. The first species has to keep evolving new features just to maintain its current chance of avoiding extinction. This reminded Van Valen of something said by the Red Queen in Lewis Carroll's *Through the Looking-Glass*: "It takes all the running you can do, to keep in the same place." He suggested that species are effectively evolving to stay where they are.

The idea has become popular with biologists, but it is not without its critics. In the late 1990s, **Anthony Barnosky** suggested that evolution is often driven by physical, non-biological pressures – changes in climate, for example – rather than by competition between species. As such, the **Red Queen's hypothesis** might not always hold true. Barnosky called this alternative reading of evolutionary change the **Court Jester hypothesis**, perhaps a reference to the surprising or disruptive role sometimes played by the joker in a deck of cards.

2/Shortcut: Van Valen's Red Queen is about the survival of species, but it is easier to think of the idea in terms of the survival of populations. Imagine a population of herbivores living alongside a population of predators. Every year the predators kill roughly one thousand herbivores. The herbivores evolve traits that reduce their vulnerability: thicker skin and faster bodies, for instance. In principle the kill rate should fall. But the predators are evolving traits that boost their chances of making a kill: stronger jaws and sharper eyes, for example. Because both populations are constantly evolving, neither ever gains the upper hand. The predators still make roughly one thousand kills every year.

Surprisingly, Alice's Red Queen is an important figure in evolutionary biology.

See also //
2 The principle of coevolution, p.8

It takes all the running you can do, to keep in the same place.

3/Hack: There's a difference between relative and absolute success: a species in an ecosystem is evolving traits that increase its chances of survival. But so are the other species in its environment.

Relatively speaking, survival rates don't change.

No.9
The theory of punctuated equilibrium
Evolution gets a radical rethink

1/Helicopter view: At the very core of **Charles Darwin**'s thinking on evolution is the idea that species evolve gradually. In the 1970s, **Stephen Jay Gould** and **Niles Eldredge** dared to suggest an alternative view. They argued that the evolution of new species is more often than not abrupt and dramatic.

Gould and Eldredge's idea really began a few decades earlier with the work of **Ernst Mayr**. In his research, Mayr suggested that the flow of genetic information through a large population of organisms should, in principle, act as a sort of buffer that stifles innovation and prevents new species appearing. However, the situation changes if a few individuals become isolated from the main population. Because this satellite population is small, useful genetic innovations are more likely to be influential rather than to be drowned out: the population is more likely to evolve into a new species.

Gould and Eldredge liked this idea. They had spent years studying the fossil record, and they knew that throughout geological history, new species seemed to appear suddenly, change little in appearance for thousands or even millions of

Punctuated equilibrium

Time

Gradualism

years, and then disappear abruptly. This didn't seem to be compatible with Darwin's view of species evolving gradually, but it fit well with Mayr's idea of species appearing rapidly in small, satellite populations.

Gould and Eldredge came up with a theory that combined Mayr's research into genes and their research into fossils. A new species appears suddenly from a small founder population. As it becomes more successful, its population size swells – and this stifles its ability to evolve and change its appearance. Eventually, it becomes less competitive and dies out. Gould and Eldredge called this view of evolution **punctuated equilibrium**.

Punctuated equilibrium argues new species appear rapidly, not gradually.

2/ Shortcut: Punctuated equilibrium is about the power of the small. Imagine being handed a pint of milk, an almost empty bottle of red food dye, and a written command to make bright pink milk. At best, the few drops of dye in the bottle will give the pint of milk a slightly pinkish hue. But by decanting a little of the milk into the dye bottle you can meet the challenge: the small volume of milk in the dye bottle is bright pink. Punctuated equilibrium argues species change only when a small subset of the population is "decanted" so it can be influenced by a new genetic mutation. For most of the time, populations are so large that they are difficult to influence – and species don't seem to evolve very much.

See also //

36 The theory of catastrophism, p.76

Stephen Jay Gould // 1941-2002

3/ Hack: A common assumption is that evolution is all about change.

The theory of punctuated equilibrium emphasizes that it's also about stasis.

No.10

The concept of exaptation

Why half a wing is more useful than it might appear

St. George Jackson Mivart //
1827–1900

1/Helicopter view: About a decade after **Charles Darwin** published *On the Origin of Species* his **theory of evolution by natural selection** was criticized by **St. George Jackson Mivart**. The problem with natural selection, argued Mivart, is that it suggests species and their features evolve very gradually. But something like an eye or a wing becomes useful only in a complete form. What good is half a wing, and why would natural selection favour an animal with such an apparently useless feature?

Darwin may actually have already addressed this apparent problem in the first edition of his book, but he discussed it more fully in the final edition published in the 1870s. The solution, he wrote, is to bear in mind that evolution acting in the distant past does not have any insight into the way animals look today. Half a wing seems useless only because we assume that wings were always "intended" to help birds fly. But if feathered wings initially appeared to fulfil some other purpose, then half a wing might well have been useful. Perhaps wings gradually grew larger because they favoured animals in some other way, until one day they were large enough that they were fortuitously suited for a brand new function: flight.

This idea came to be known as preadaptation, but biologists hated the term. It seemed to imply that evolution is, in some way, goal-directed: that it "preadapts" species in preparation for a specific challenge ahead. Recognizing this semantic problem, **Stephen Jay Gould** and **Elisabeth Vrba** suggested in 1982 that biologists might want to use the term **exaptation** instead of preadaptation. The scientific community agreed.

Archaeopteryx (main image) and *Sinornithosaurus* (inset) reveal stages in the evolution of feathers.

2/ Shortcut: Gould and Vrba argued that species are constantly adapting and exapting. For instance, long ago proto-birds may have developed short feathers to trap heat and keep their bodies warm – an adaptation. Eventually, feathers evolved to be so long that they may have given the proto-birds an unexpected benefit: they became impromptu nets for catching insects for food – an exaptation. Natural selection then favoured animals with more elaborate feathers that were even better insect nets – an adaptation. Eventually, feathers became so elaborate that the proto-birds gained another unexpected benefit: the feathers could help them glide short distances – an exaptation. And so on…

See also //

1 The theory of evolution by natural selection, p.6

15 The process of genetic assimilation, p.34

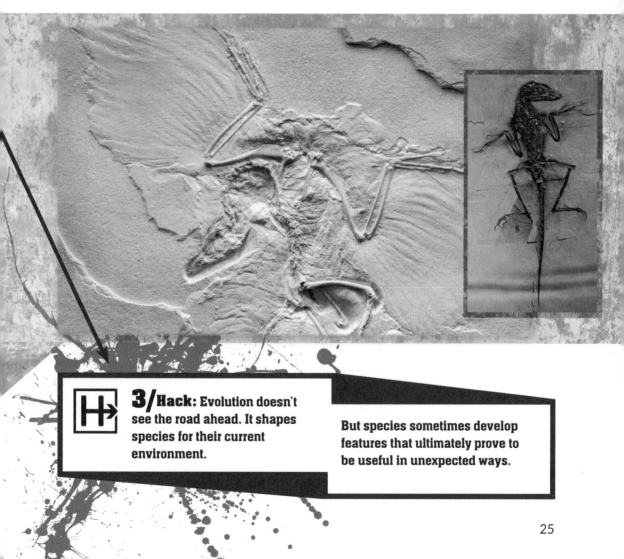

3/ Hack: Evolution doesn't see the road ahead. It shapes species for their current environment.

But species sometimes develop features that ultimately prove to be useful in unexpected ways.

The process of convergent evolution

Déjà vu all over again

Charles Darwin // 1809–1882

1/ Helicopter view: In 1859 **Charles Darwin** wrote in great detail about the evidence he had amassed in favour of the idea that evolution occurred through natural selection. The final sentence of his famous book, *On the Origin of Species*, suggests that this relatively simple process has produced "endless forms most beautiful and most wonderful". There is just one problem with this conclusion. Evolution doesn't seem to produce "endless" variety.

In reality, the natural world appears to be rife with repetition. Often, organisms independently evolve the same "solution" to a given ecological "problem" – and consequently they end up looking rather similar to each other. The process has come to be known as **convergent evolution**. In 2011, **George McGhee** suggested tweaking Darwin's wording to acknowledge that today's ecosystems and the fossil record actually seem to record evidence of "*limited* forms most beautiful".

Genetics might hold the key to understanding why convergent evolution seems to occur so often. Naively, we might assume the DNA mutations that help drive evolution are as likely to occur anywhere in the genome. In fact, some regions of the genome seem to be more likely to mutate than others.

What's more, DNA mutations are not "born equal". Some may alter the way a gene operates in every single cell in the body, which is such a radical change it might actually kill the organism. Other mutations influence the way genes operate in just some of an organism's cells. These mutations are less drastic and more likely to benefit the organism in a way that could encourage evolution to occur (see: **The junk DNA question**, page 52). Partly due to constraints like these, evolution might not actually be completely random and unpredictable after all.

The ancient ichthyosaurs looked remarkably like today's dolphins.

2/Shortcut: The history of life on Earth is full of examples of convergent evolution. For instance, hundreds of millions of years ago a group of land-living reptiles gradually evolved to become formidable marine predators – ichthyosaurs. A few tens of millions of years ago a group of land-living mammals also evolved to become marine predators – dolphins. To the untrained eye, ichthyosaurs and dolphins look very similar, even though they evolved completely independently from different ancestors.

See also //

24 The junk DNA question, p. 52

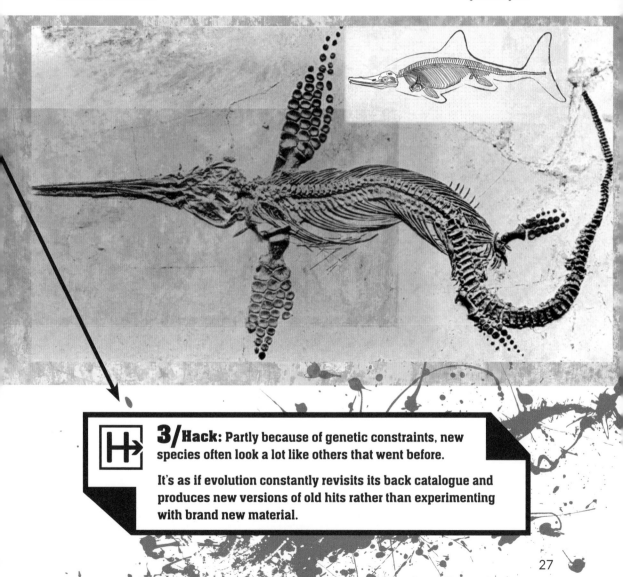

3/Hack: Partly because of genetic constraints, new species often look a lot like others that went before.

It's as if evolution constantly revisits its back catalogue and produces new versions of old hits rather than experimenting with brand new material.

No.12
Horizontal gene transfer

Evolution but not as Darwin predicted it

Oswald Avery // 1877–1955

1/Helicopter view: In 1928 **Frederick Griffith** made an astonishing and disturbing discovery. He was investigating two strains of a species of bacteria: strain A was harmless when injected into lab mice, but strain B triggered lethal pneumonia in the rodents. He found that blasting the lethal strain B bacteria with heat was enough to kill the microbes, rendering them as harmless as strain A. Then came the unexpected find: he mixed living bacteria from strain A with dead bacteria from strain B and injected them into mice. Individually, both strains were harmless – but together, they killed the mice. Griffith's work was the first evidence that evolution can work in a completely unexpected way.

Griffith concluded that the dead bacteria had somehow passed their "killing potential" to the living microbes. But this was a hugely controversial suggestion. Scientists going back to **Charles Darwin** had assumed that characteristics pass exclusively between (living) parents and their offspring – called "vertical" inheritance. Killing the lethal bacteria rendered them unable to reproduce. Their power to kill animals should have died with them.

Oswald Avery read Griffith's results and decided to investigate. In 1944 he made a key discovery: he identified a molecule that passed between the dead and the living bacteria, presumably carrying the "lethal" trait with it. The molecule was DNA. Avery's was a landmark study: not only did it help explain Griffith's result, it also helped geneticists establish that DNA was the main molecule involved in inheritance, which would have far-reaching implications for scientific study later in the 20th century.

Just as importantly, Avery and Griffith had collectively shown that inheritance is not always strictly vertical between parents and offspring. Sometimes genetic material can pass "horizontally" between completely unrelated organisms that just happen to inhabit the same environment. No one, not even Darwin, had predicted the **process of horizontal gene transfer**.

Bacterial cells in a culture can swap packets of DNA with one another.

2/Shortcut: We are familiar with vertical inheritance: human children inherit genes, and the physical features they code for, from their parents. Microbes behave differently. Many of them can produce a small packet containing a copy of some of their DNA, which they then pass to another microbe they meet, whether or not they are closely related. The extent to which this horizontal gene transfer occurs among animals as well as just microbes is still being discussed.

See also //

19 The double helix model, p.42

50 The concept of antibiotic resistance, p.104

3/Hack: Genetic material is a bit like a piece of juicy gossip: sometimes it is too good to keep within the family.

Instead it spreads "horizontally" to strangers in the community.

No.13
Endosymbiotic
theory

How teamwork gave the world complex life

1/Helicopter view: The quality of microscopes improved substantially during the 19th century, allowing biologists to peer at, and inside, biological cells. In 1883, **Andreas Schimper** noticed that tiny structures called chloroplasts inside plant cells looked strangely familiar. In fact, they looked a lot like special strains of bacteria that can produce their own food through photosynthesis. Schimper speculated that plants might have originated with some sort of strange biological union involving bacteria. It seemed like a mad idea, but within a century it had become scientific consensus.

In the early part of the 20th century, **Konstantin Mereschkowski**, aware of Schimper's observations, fleshed out the idea in more detail. By now biologists knew that two unrelated organisms sometimes enter into a mutually beneficial (or "symbiotic") relationship. Mereschkowski argued that an extreme version of this had happened in plants. He imagined (probably incorrectly) that life on Earth had evolved twice: one form of life became bacteria and the other became slightly more complicated amoeba-like cells. At some point, the two forms of life met: an amoeba joined forces with a photosynthetic bacterial cell and the two became a single-celled alga, the distant ancestor of all plants. Biologists ignored the idea for several decades.

Andreas Schimper //
1856–1901

By the 1950s, geneticists had begun to accept that DNA is the key molecule organisms use to replicate themselves. They made an unexpected discovery: the chloroplasts inside plant cells have their own packet of DNA operating independently of the plant's DNA. The finding hinted that Mereschkowski was correct. Chloroplasts were once independent free-living microbes using their DNA to replicate themselves, even though today they cannot function without plant cells (and plant cells cannot function without chloroplasts). At roughly the same time, biologists discovered that small structures called mitochondria inside plant, animal and fungal cells also contained their own DNA. In 1967, **Lynn Margulis** revived Mereschkowski's idea. The **endosymbiotic theory** was on the road to widespread acceptance.

Chloroplasts inside plant cells might once have been free-living microbes.

2/Shortcut: Margulis's endosymbiotic theory suggests that, deep in prehistory, a dinner date went awry. One microbe swallowed another, but the smaller cell somehow escaped being digested. Instead, it entered into a beneficial relationship with the larger cell. The larger cell protected the smaller cell and provided it with nutrients, while the smaller cell produced lots of energy for the larger cell. Eventually the two cells functioned as one complex organism – an organism that evolved into animals and fungi. Its descendants also evolved into plants – but only after another hapless dinner resulted in the addition of a third microbe to the system, this one ultimately becoming the photosynthetic chloroplast.

See also //
14 The LUCA hypothesis, p.32

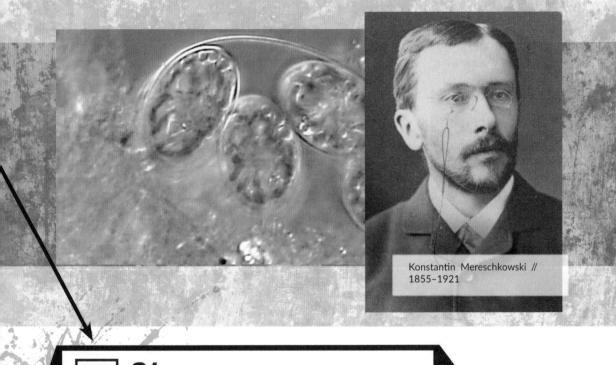

Konstantin Mereschkowski // 1855–1921

3/Hack: Partnerships come and go, but according to endosymbiotic theory some of the most intimate ones have lasted a very long time.

These microbial partnerships are responsible for all complex life on Earth.

No.14
The LUCA hypothesis
The ancient microbe with an astonishing legacy

1/Helicopter view: In 1859, **Charles Darwin** published one of the most famous scientific books of all time: *On the Origin of Species*. In it, Darwin set out his **theory of evolution by natural selection**, describing a process through which new species might appear on Earth. Darwin also made a bold prediction: he speculated that all living things might ultimately be descendants of a few – or even a single – species that lived long, long ago. Today, scientists suspect he was correct.

Long after Darwin lived, scientists began to recognize that all cellular life (everything from bacteria to elephants) shares a common feature: a genetic code built using a molecule called DNA. Fundamentally, that DNA is very similar in all cellular life forms, which strongly suggests all living species inherited their DNA from the same ancestor – a hypothetical species that scientists call the **Last Universal Common Ancestor,** or **LUCA.**

LUCA was not the earliest species on Earth. In fact, by the time it lived our planet was probably populated with a huge array of species. But through a quirk of fate the descendants of all the species that shared LUCA's world ultimately disappeared.

It is impossible to say exactly what LUCA looked like, or when and where it lived. This is because comparatively few species leave traces in the fossil record, making it very unlikely that fossil evidence of LUCA will ever be found. But many biologists suspect that LUCA was a single-celled microbe living roughly 3.8 billion years ago. Some genetic studies suggest LUCA lived around hydrothermal vents on the seafloor.

LUCA

All life on Earth today might have stemmed from an ancient microbe living around deep sea vents.

2/ Shortcut: The LUCA hypothesis is probably best understood through analogy. Imagine 100 saplings, each with just a single branch, planted together on a small patch of ground. As the years pass, the saplings grow and send out new branches. Some grow more vigorously than others – weaker individuals are choked out and die. After a century only one individual still survives, and it has become a magnificent tree with tens of thousands of branches in its crown. Think of this as the tree of life, with each of its branches representing one of the many species alive on Earth today. We can think of the single-branched sapling that grew into this tree as LUCA. It wasn't the only sapling alive to begin with, but eventually only the branches that it grew survived.

See also //

1 Theory of evolution by natural selection, p.6

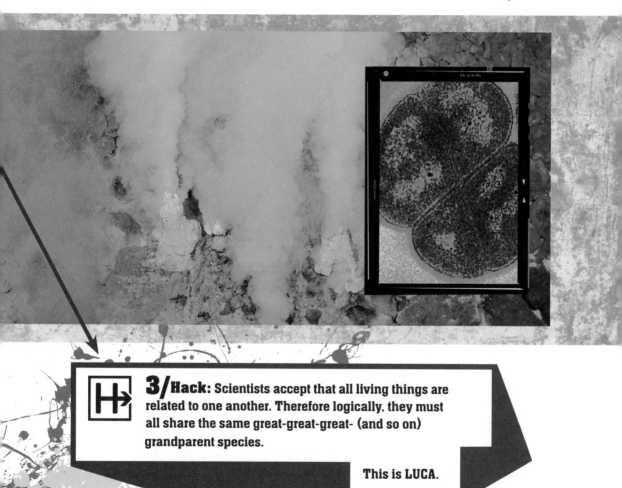

3/ Hack: Scientists accept that all living things are related to one another. Therefore logically, they must all share the same great-great-great- (and so on) grandparent species.

This is LUCA.

No.15

Conrad Hal Waddington //
1905–1975

The process of genetic assimilation

Evolution back to front

 1/ Helicopter view: Early in the 1950s, **Conrad Hal Waddington** made a curious discovery. He was experimenting with fruit flies, investigating the fact that the insects develop unusual wings if they are grown at an abnormally warm temperature. After growing several consecutive generations in warmer conditions, Waddington turned down the temperature. The flies continued to develop the unusual wings. Waddington had triggered evolution – but not in the way that most biologists think it should happen.

Waddington's results came at a pivotal time in the history of biology. Just a few years earlier scientists had first realized that DNA is the molecule responsible for inheritance. The origin of new species is, fundamentally, about inheritance. As such, it seemed obvious to link the appearance of new species to the acquisition of new (heritable) DNA mutations. The fruit fly experiment did not fit with this narrative.

Biologists had actually known for decades that organisms can change their appearance and behaviour dramatically if the environment they occupy changes. The process is called **developmental plasticity.** However, because developmental plasticity does not involve mutating DNA it seemed unlikely that it could lead to the heritable changes that drive evolution. And yet, Waddington's experiments hinted that developmental plasticity *had* triggered evolution in the fruit flies.

Waddington called this effect **genetic assimilation**. In the 1950s, it was largely dismissed as a curiosity – but not today. **Kevin Laland** and others think genetic assimilation might play a key role in the evolution of new species. These scientists have put the idea at the heart of a new "extended evolutionary synthesis" which challenges central assumptions of the gene-centric view of evolution that has dominated for almost a century. Evolutionary thinking is continuing to evolve.

Tiktaalik, a 375-million-year-old extinct species, was a fish with leg-like fins.

2/ Shortcut: Genetic assimilation suggests evolution can occur in an odd way. Waddington's fruit flies developed unusual wings because they were raised in warm conditions. The change is not hereditary, but because Waddington raised several "warm" generations all of the flies developed the same unusual wings. As the generations passed the flies' DNA naturally picked up new mutations. Some made life easier for the flies with the unusual wings and so these mutations became more common in the population. By the time Waddington turned down the temperature, several of these (heritable) mutations were ready to act as a genetic scaffold constraining the flies to continue developing the unusual wings. Recent experiments hint that genetic assimilation might have been at work millions of years ago when fish first crawled onto land and their fins became legs.

See also //

4 The concept of neoteny, p.12

5 The modern synthesis, p.14

3/ Hack: Most biologists assume changes to DNA drive the origin of new species — genetic assimiliation suggests an alternative.

It may be non-heritable flexibility in the way an organism grows to adulthood that comes first.

The theory of Mendelian inheritance
Peas behaving oddly

Gregor Mendel //
1822–1884

1/Helicopter view: In the middle of the 19th century **Gregor Mendel** used the garden of his Austrian monastery as the setting for some fairly basic hybridization experiments – and made a discovery that helped kickstart a new field of scientific exploration: genetics.

Mendel took pea plants with white flowers and crossed them with a purple-flowered strain. Unexpectedly, the hybrid seeds did not grow into plants with purplish-white flowers. Instead, they all had solidly purple blooms. Mendel next fertilized these hybrid purple-flowered peas with one another to produce a new batch of seeds. Three-quarters of the plants these seeds grew into also had vibrant purple flowers – but one-quarter had pure white flowers, just like half of the pea plants that Mendel had begun with.

Mendel realized there was something unusual going on, and he came up with a **theory of inheritance** to account for the results. Put simply, he suggested that characteristics like flower colour are governed by discrete chunks of information called "factors" (better known today as genes) that are heritable. Mendel reasoned that, in at least some situations, a biological characteristic must be controlled by two versions of a factor (what are now called alleles). He further suggested that one of these two versions is dominant, meaning that if it is present it governs how the biological characteristic manifests in the final organism.

With simple ideas like this Mendel could explain the odd results of his plant experiments. Unfortunately for Mendel, the scientists of his day largely dismissed his findings. The prevailing view was that inheritance involved a general blending, or smearing, of the characteristics of both parents – although this view hadn't been formalized in a theory. Mendel suspected the blending idea was wrong, given that his pea plants always had *either* purple flowers *or* white flowers but never purplish-white flowers – but it would be several decades before biologists accepted his idea.

The science of genetics began with pea plants.

2/Shortcut: Mendel began with two purebred pea strains. One had purple flowers because it carried two dominant versions of the "flower colour" allele – PP. The other had white flowers because it carried two recessive versions – pp. Logically, when these two plants hybridize their offspring inherit one allele from each parent. Genetically, they are all Pp and will produce purple flowers, because they all carry a copy of the dominant "flower colour" P allele. When Pp flowers are crossed, four genetic combinations are possible: PP, Pp, pP and pp. Three of the four carry a dominant allele (P) so will produce purple flowers. One carries only recessive alleles (pp) so will produce white flowers.

See also //

5 The modern synthesis, p.14

21 Selfish gene theory, p.46

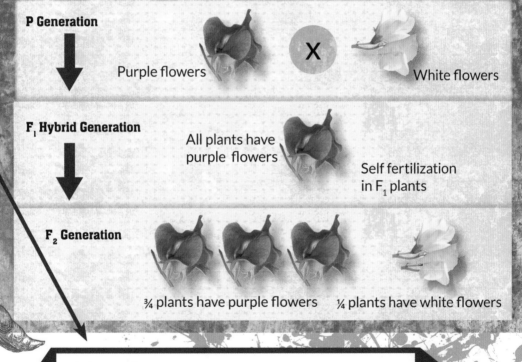

P Generation

Purple flowers X White flowers

F₁ Hybrid Generation

All plants have purple flowers

Self fertilization in F₁ plants

F₂ Generation

¾ plants have purple flowers   ¼ plants have white flowers

3/Hack: Gregor Mendel is credited as being the first to realize that parents pass discrete chunks of information to their offspring.

In effect, he recognized the existence of genes.

No.17
Chromosome theory of inheritance

How genetics turned mainstream

 1/ Helicopter view: In the 1860s **Gregor Mendel** suggested that offspring inherit discrete chunks of information (what scientists now call genes) from their parents – but even 40 years later few scientists took his idea seriously. Not least of the problems was the lack of an obvious mechanism through which an individual could gain half of its genes from one parent and half from the other, as Mendel's idea seemed to require. Eventually, scientists realized that chromosomes provided that mechanism.

The 19th century had brought an improvement in the quality of microscopes, allowing scientists to confirm that many cells contain a small internal structure: the cell nucleus. Further observations showed that just before and immediately after a cell divides, mysterious strands become visible in the nucleus. These strands became known as chromosomes.

In the 1880s, **Theodor Boveri** began to study chromosomes, convinced that they must play some role in the inheritance process. Some of his most important discoveries involved sea urchin eggs. These eggs can sometimes be fertilized by two sperm cells rather than the standard one: Boveri found that such fertilized eggs often carry more than the 36 chromosomes that are standard in healthy sea urchin cells. When they did, though, the egg failed to develop properly. This suggested that healthy embryo development occurs only when the fertilized egg carries an equal number of chromosomes from both parents.

A few years later Mendel's papers on inheritance were rediscovered. Boveri realized that his research into chromosomes was compatible with Mendel's ideas. In particular, he reasoned that Mendel's "factors" (genes) of inheritance might lie on the chromosomes. Another scientist, **Walter Sutton**, was arriving at a similar conclusion at roughly the same time.

In 1915, **Thomas Hunt Morgan** put all the pieces of the puzzle together in an influential book, and the study of genes moved from the scientific margins to the mainstream.

Theodor Boveri //
1862–1915

Chromosomes condense and become visible under a standard microscope when cells divide.

2/ Shortcut: Inside their cells, organisms carry two copies of each chromosome. Given that genes lie along the chromosome, this means organisms have two copies of each gene, as Mendel suspected. But something odd happens during sexual reproduction. Sperm and egg cells carry just one copy of each chromosome. When they combine during conception, the resulting fertilized egg carries one copy of any given gene from the mother and one copy from the father, just as Mendel had theorized.

See also //

16 The theory of Mendelian inheritance, p.36

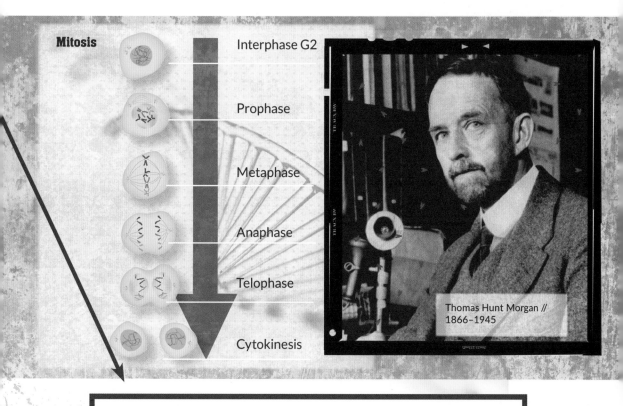

Mitosis

Interphase G2

Prophase

Metaphase

Anaphase

Telophase

Cytokinesis

Thomas Hunt Morgan //
1866–1945

3/ Hack: The genes in chromosomes become more tightly packed as cells divide, which means the chromosomes themselves become visible through a standard microscope.

Studying chromosome behaviour gave biologists their first insight into the mechanics of genetic inheritance.

No.18

The process of genetic drift

The role luck plays in evolution

1/ Helicopter view: In the 1920s biologists began to combine **Charles Darwin**'s ideas on natural selection with **Gregor Mendel**'s ideas on inheritance. But in doing so they discovered that the process of evolution might be even more random than many had anticipated.

Sewall Wright // 1889–1988

The problem began when **Sewall Wright** considered how populations of organisms might behave at the genetic level. Most of Wright's colleagues assumed that the individuals that had the most offspring would be those with characteristics (and genes) best suited for the environment. Wright argued that simple random chance might be involved too.

He thought this was particularly likely to occur in small populations where an individual can have more of an impact on the way the population evolves in future. Supposing one organism "gets lucky" and has twice as many offspring as would be expected from the quality of its genes. Its DNA will be more common in the next generation not because it is beneficial to the species, but simply down to random chance. If this keeps happening generation after generation, it might influence the evolution of the population, and the species, as a whole. Wright called this process **genetic drift**.

Wright's work sparked a heated debate. Some of his peers, including **Ronald Fisher**, accepted that genetic drift might occur but were adamant that evolution was largely driven by natural selection. But Wright's idea received a boost in the 1960s. **Motoo Kimura** argued on mathematical grounds that random drift must be hugely important in evolution. To this day, scientists have not reached a consensus about the relative importance of genetic drift and natural selection in driving evolution.

Random fortune might consign some species to extinction.

2/ Shortcut: Genetic drift suggests evolution is even more random than Darwin suspected. Imagine a population of animals in which a gene mutates in two ways: one that benefits its carriers, and one that makes life more challenging for its carriers. Logically, the beneficial mutation will become more common and the detrimental one less so. But by sheer chance, all "beneficial" carriers might one day be clustered in a valley that experiences a natural disaster – a lava flow. A few "detrimental" carriers survive, and as the population recovers, it's their genes that become prevalent. The detrimental gene "won", not because it gave its carriers an advantage (it didn't) but simply through lucky circumstance.

See also //

5 The modern synthesis, p.14

9 The theory of punctuated equilibrium, p.22

3/ Hack: Some individuals might have a successful sex life (and pass on their genes) because of random good fortune rather than because they were born with beneficial adaptations.

This is survival of the luckiest rather than the fittest.

No.19
The double helix model
DNA finally makes sense

1/ Helicopter view: In the early 1950s several scientists were unofficially involved in a frantic race to work out what a molecule of DNA actually looks like. Winning that race made two of those scientists – **Francis Crick** and **James Watson** – world-famous.

The study of DNA stretches back into the 19th century. **Friedrich Miescher** identified the molecule just a decade after **Charles Darwin** published *On the Origin of Species* – although the significance of Miescher's discovery was unappreciated for decades. This wasn't surprising. Few scientists suspected that DNA was involved in inheritance. To put it another way, few thought that genes were coded in DNA.

Oswald Avery surprised many when, in 1944, he provided strong evidence that genes are indeed built from DNA. Suddenly it dawned on scientists that they needed to work out the precise structure of DNA to fully understand inheritance.

Erwin Chargaff made an important contribution in 1950: he found that structures in DNA called "bases" were balanced. In any given DNA molecule, the amount of a base called adenine roughly equalled the amount of a base called thymine, while the amount of guanine roughly equalled the amount of cytosine.

James Watson // b. 1928

Within three years, Crick and Watson had come up with a model for the structure of DNA that explained Chargaff's observation. Borrowing heavily from research performed by **Rosalind Franklin**, **Maurice Wilkins** and **Linus Pauling**, Crick and Watson came up with the famous **double helix model of DNA**. The molecule at the heart of evolution finally made sense.

Three of the key figures in the race to unravel DNA's structure.

2/ Shortcut: Crick and Watson's double helix model for DNA is one of the most famous of all scientific images. That's not simply because it looks elegant. The double helix also explains how genes are replicated and passed down the generations. The model is like a twisted ladder, with DNA's "bases" combining in pairs to form the ladder's rungs – adenine always pairing with thymine (A-T) and cytosine pairing with guanine (C-G). The important point is that if DNA "unzips" along its length, breaking up each A-T and C-G base-pair, the two halves of the molecule can act as a scaffold on which the missing half can be built. DNA's structure is ideally suited for self-replication.

See also //

12 Horizontal gene transfer, p.28

Francis Crick // 1916–2004

Rosalind Franklin // 1920–1958

3/ Hack: Crick and Watson did not discover DNA. But their double helix model did solve a mystery that stretched back to Darwin.

They showed how DNA can copy itself, providing a mechanism for inheritance.

43

No.20
The molecular clock hypothesis
Measuring evolutionary time

1/Helicopter view: In 1960, **Linus Pauling** was invited to submit a paper to a special volume celebrating the work of **Albert Szent-Györgyi**, who had discovered vitamin C in the 1930s. In the paper, Pauling and his colleague, **Emile Zuckerkandl**, came up with an idea that revolutionized the evolutionary sciences.

Pauling and Zuckerkandl revised an idea from the beginning of the 20th century that complex molecules (proteins and DNA) inside biological cells are gradually, but constantly, changing their appearance. The two scientists realized that these molecules were, in effect, timepieces that could be used to work out when two cells – or, more usefully, two species – last shared a common ancestor.

This was an extraordinary and controversial idea. At the time most scientists thought the fossil record alone could help establish exactly when in prehistory new species had first appeared. Pauling and Zuckerkandl dared to suggest that molecules inside living organisms could do the job as well. Even worse, when the two scientists used their idea to calculate when humans and gorillas diverged from a common ancestor the answer they came up with – eleven million years ago – disagreed with the fossil record as it was understood at the time.

As the years passed, though, more fossils were found and Pauling and Zuckerkandl's eleven million year figure began to seem remarkably plausible. The **molecular clock hypothesis** had passed a crucial test: it was here to stay.

Linus Pauling //
1901–1994

Two species separated by many genetic mutations (nucleotide substitutions) last shared a common ancestor long ago.

2/Shortcut: Complex organic molecules like DNA are fragile. When cells divide – and DNA replicates – it almost inevitably accumulates tiny errors. Like a game of Chinese whispers or telephone, more and more errors accumulate with every additional replication. Pauling and Zuckerkandl's molecular clock works on the assumption that the rate these errors accumulate is roughly constant. This means that if two species last shared a common ancestor ten million years ago, comparing their DNA should reveal twice as many errors as would be seen when comparing the DNA of two species that last shared a common ancestor just five million years ago.

See also //

22 The mitochondrial Eve hypothesis, p.48

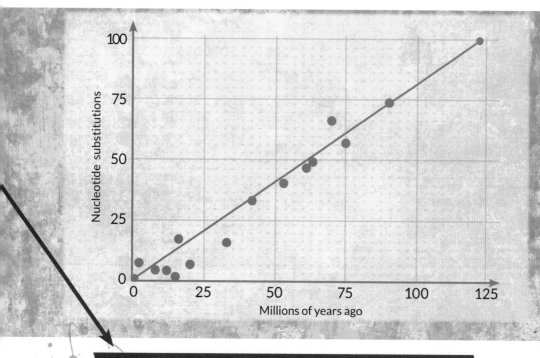

3/Hack: Scientists once thought that only fossils could reveal when in prehistory species diverged from each other.

The molecular clock allows biologists to use living animals as an alternative tool for exploring prehistoric time.

No.21

Richard Dawkins // b. 1941

Selfish gene theory
A new perspective on evolution

1/Helicopter view: In 1976, **Richard Dawkins** published a book that would, like **Charles Darwin**'s *On the Origin of Species*, appeal both to a general and a scientific readership. And just as Darwin's book influenced the future direction of scientific investigation, Dawkins's did too. The book in question is *The Selfish Gene*.

The Selfish Gene was the end result of a scientific journey that arguably began in the 1860s with **Gregor Mendel**'s work on inheritance. By the 1930s, scientists had recognized the evolutionary importance of Mendel's work – which identified that units of information (genes) pass essentially unchanged down the generations.

Further key breakthroughs came in the 1940s and 1950s. Geneticists established that DNA is the molecule responsible for inheritance, strongly suggesting genes were coded in DNA. They also worked out exactly how DNA can replicate itself faithfully when new cells or new organisms are generated.

All of these discoveries led scientists including **George Williams** and **John Maynard Smith** to argue for a gene-centred view of evolution. At a fundamental level, they said, evolution is about the survival of genes, not of species – and certainly not of individual organisms. Over millions of years of evolution individuals die, species become extinct – but genes persist. This way of thinking helped explain some biological oddities including altruistic behaviour and it became the central theme of Dawkins's famous book. The **selfish gene theory** that emerged from the book was – and is – highly influential.

An athlete in a kayak provides an easy way to understand the "selfish" nature of genes.

46

2/ Shortcut: Borrowing one of Dawkins's analogies, imagine an Olympic kayaker's career represents the history of life on Earth. As the kayaker trains and improves her skills she uses many different kayaks. Some of them break up from overuse on the river rapids, others are replaced with newer, better versions. Sportswriters recognize that these various kayaks are important, but they know it is what is inside each kayak that really counts: the kayaker herself. Similarly, when documenting the history of life on Earth, it is not individuals and species that really matter – it is the genes that move from body to body.

See also //

7 The concept of kin selection, p.18

16 The theory of Mendelian inheritance, p.36

19 The double helix model, p.42

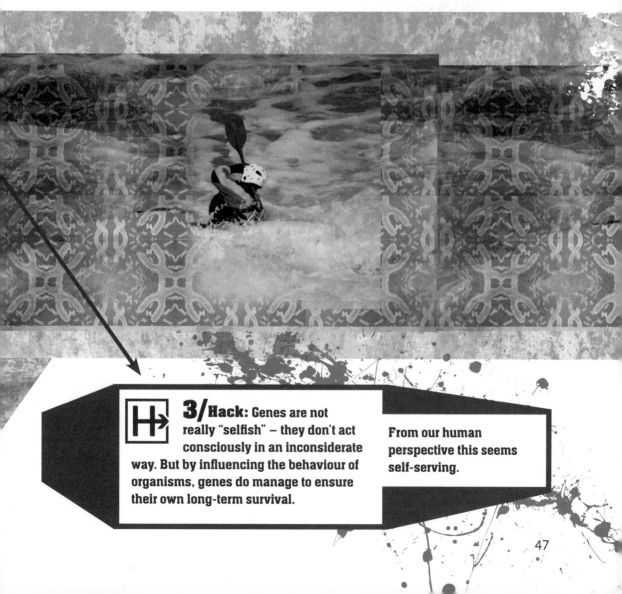

3/ Hack: Genes are not really "selfish" – they don't act consciously in an inconsiderate way. But by influencing the behaviour of organisms, genes do manage to ensure their own long-term survival.

From our human perspective this seems self-serving.

No.22
The mitochondrial Eve hypothesis

The "mother" of humanity

1/ Helicopter view: In 1987 three biologists – **Rebecca Cann**, **Mark Stoneking** and **Allan Wilson** – published a paper with implications that captured the public imagination. They had analysed samples of human DNA and made the bold suggestion that every man, woman and child alive today inherited one particular chunk of DNA from a single common ancestor: a woman who lived long ago in Africa. The newspapers loved the idea. They dubbed the hypothetical ancient woman "Eve".

Rebecca Cann //
b. 1951

The discovery of this particular Eve followed on from at least two earlier scientific breakthroughs. First came the discovery, in the 1950s, that small structures in human cells called mitochondria carry their own tiny chunk of DNA that operates independently of the human genome. Mitochondrial DNA behaves in an odd way: when a child is born it inherits half of its human genome from its mother and half from its father. However, a child's mitochondrial DNA always comes exclusively from the mother.

The second important discovery, made in the 1960s, is that biological molecules gradually and constantly change their appearance from generation to generation, meaning that geneticists can work out when two people last shared a common ancestor simply by counting the small genetic differences in their DNA.

The three scientists analysed mitochondrial DNA samples volunteered by 147 women across the world and constructed a grand, global family tree. They discovered that this tree traced its roots back to Africa, some 140,000 to 290,000 years ago. At some point in that time window, there lived a woman who through random fortune ultimately "gave" her mitochondrial DNA to everyone alive today. Wilson liked to call this woman the "lucky mother". But partly because of the media reporting in 1987, the idea is more widely known by a more evocative name: the **mitochondrial Eve hypothesis**.

Mitochondria (in pale blue) carry their own DNA – and their own story about humanity's past.

2/ Shortcut: Mitochondrial Eve is easiest to understand through analogy. Imagine that your best friend and you represent the entire global human population. You each carry DNA from four grandparents, from eight great grandparents, and so on – plenty of ancestors. However, mitochondrial DNA is unique because it passes exclusively down the maternal line: this means you each carry mitochondrial DNA from just one grandmother, from one great grandmother, and so on. Go far enough back down your maternal line – and far enough back down your friend's – and eventually you will find a woman common to both of them. Neither you nor your friend trace *all* of your DNA to this woman – you each have scores of other great, great, great (and so on) grandmothers and grandfathers. But the woman did ultimately pass her *mitochondrial* DNA to both you and your friend. She is your shared mitochondrial Eve.

See also //

13 Endosymbiotic theory, p.30

20 The molecular clock hypothesis, p.44

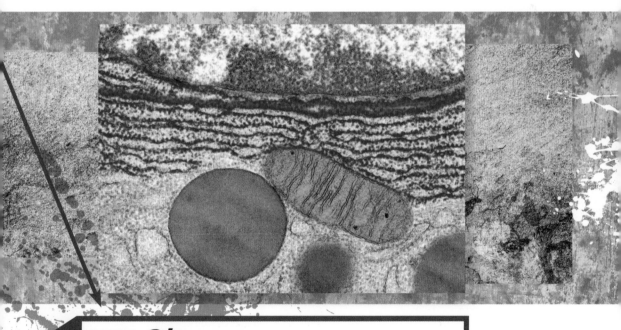

3/ Hack: For all our differences, people really do belong to the same global family with the same distant ancestors. Mitochondrial Eve embodies this idea.

She is a prehistoric individual who gave some of her DNA to every person alive today.

No.23
The RNA world
hypothesis
The dawn of genetics?

1/ Helicopter view: By the second half of the 20th century, scientists had begun to realize that the system all living things use to replicate themselves is incredibly elaborate. It involves three different types of complex organic molecule: DNA, RNA and proteins. This left them with a puzzle: how could such an intricate and complicated system have come into being in the first place?

In the 1960s, **Leslie Orgel** and **Francis Crick**, along with a few other biologists, argued it must have had a simpler precursor. At the dawn of life on Earth there must have been a system in which just one type of complex molecule was involved in replication. They suspected that this molecule was RNA.

Today, RNA seems to play a relatively minor role in replication. The genetic code for constructing a new organism is stored in DNA. This code is used to make biological "bricks" – proteins – that actually build the organism. RNA seems to be little more than a courier service that carries genetic messages between the DNA and the cellular machines where the proteins are built. But Orgel and Crick suspected that, at the very beginning of life on Earth, RNA did much more.

Specifically, the researchers argued that RNA would ultimately turn out to be capable of functioning both as a genetic store – like DNA – and as a biological building block – like proteins. The first half of their argument had already been confirmed. It was clear that RNA could carry genetic information, since it ferries genetic instructions between DNA and the protein-building machines.

Many years later, in the early 1980s, **Thomas Cech** confirmed the second part of Orgel and Crick's argument. He found that RNA could, indeed, behave like a protein.

Although this is not, on its own, enough to convince the entire scientific community that life began with RNA, the idea is still one of the leading hypotheses. In 1986, in a review article, **Walter Gilbert** gave it a catchy name. Ever since, the idea that the first living things functioned using only RNA has been called the **RNA world hypothesis**.

The molecular structure of RNA suggests it could have helped the earliest living things build and replicate.

2/Shortcut: The DNA molecule is like a tiny twisted ladder. The genetic code itself lies on the ladder's rungs. Proteins are very different: they are large molecular tangles – and it is their ability to fold into complex three-dimensional shapes that gives them their building block properties. RNA has both of these features. It looks a little like half of a DNA molecule, so it still carries "rungs" on which genetic information can be stored. But because those rungs are attached to only one "pole" rather than two, the entire RNA molecule is far more flexible than DNA. This means it can fold into complex three-dimensional shapes just like proteins can. To put it another way, RNA can act a bit like DNA and a bit like proteins.

See also //
14 The LUCA hypothesis, p.32
100 The panspermia hypothesis, p.204

Walter Gilbert // b. 1932

3/Hack: Building a new organism involves a complex system with three distinct types of very specialized molecules (DNA, RNA and proteins).

But perhaps the very first living things replicated themselves using just RNA jacks-of-all-trades.

No.24

The Junk DNA question
What's really going on in our cells?

1/Helicopter view: By the early 1970s geneticists had more or less worked out the way organisms build and reproduce themselves using DNA. But there were still mysteries to unravel.

For instance, **CA Thomas, Jr** was confused by the fact that relatively simple organisms sometimes contained much more DNA than relatively complex ones. He called this the C-value paradox. At roughly the same time **Susumu Ohno** popularized a controversial solution to the mystery. Perhaps genomes varied in size in a confusing way because most of the DNA inside cells did not actually serve any useful purpose at all. Perhaps it was just junk.

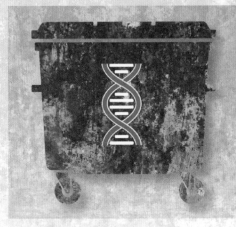

Many biologists were very unhappy with this idea. **Roy John Britten** and **David Kohne** described it as a "repugnant" concept. After all, if a large proportion of the DNA in cells really serves no purpose, then organisms should have gradually got rid of it after millions of years of evolution.

But in the early years of the 21st century it became clear that the idea might in fact be true. Geneticists had spent years carefully "reading" the human genome – the enormous DNA molecule that helps build our bodies. Many geneticists assumed that building a complicated human body would require the action of one hundred thousand different genes, each "coding" for a distinct protein. In fact, the human genome turned out to contain between twenty thousand and thirty thousand genes – and those genes collectively added up to just one or two per cent of the entire genome. It was dramatic confirmation that a huge volume of the human genome doesn't "code" for proteins. Most of our DNA is non-coding – and many biologists think a good chunk of it is simply **junk DNA**.

2/Shortcut: There is an ongoing debate in biology about the junk DNA question. Scientists can all agree that just a tiny fraction of the human genome exists in the form of genes that code for proteins, the building blocks of life. Some scientists think that most – maybe even all – of the "non-coding" DNA is still important: it might help control the way the "coding" DNA actually works in each cell. But others think a lot of non-coding DNA really is just junk. It has accumulated in our cells over millions of years, and because it doesn't "cost" our cells much to look after the junk, evolution hasn't got rid of it.

How much of the DNA in our cells is useless junk?

See also //
11 The process of convergent evolution, p.26

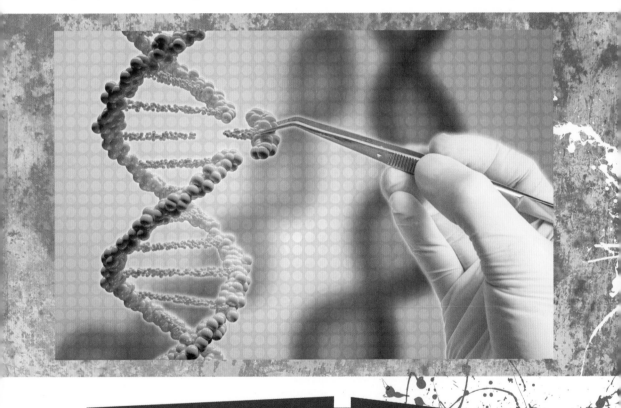

3/Hack: Evolution is pragmatic: if it is more trouble to carefully remove and throw away old useless DNA than to simply keep hold of it, genomes could well end up accumulating lots of junk.

No.25
The missing link
Why it will never be found

1/Helicopter view: In 1859, **Charles Darwin** published *On the Origin of Species*, explaining that natural selection allowed one "parent" species to evolve into two or more new "daughter" species. Initially, Darwin avoided applying his concepts of speciation to humans, possibly to avoid controversy. His colleague **Thomas Huxley** had no such qualms: in 1863, Huxley published a book that mused on humanity's place in the animal kingdom.

Huxley pointed out that, anatomically speaking, humans are very similar to chimpanzees and gorillas. He suggested all three species might have descended from the same parent species (although today the consensus view is that gorillas are slightly more distant relatives, with chimps and humans sharing a more recent parent). Scientists call this hypothetical parent species the "last common ancestor". It is also known by a less scientific term: the **missing link**.

The term "missing link" may be a hangover from a time when most people assumed there was a great chain of being incorporating all matter and life, from immortal deities at the top to rocks at the bottom. In this model, humans are higher up the chain than other animals, so a species that bridges the gap between known animals and humans must be, literally, a missing link in the chain.

It's a flawed way of thinking, though, because it nudges people towards viewing the missing link as evolving from chimps into humans. In the 1990s **Tim White** and his colleagues discovered a fossil species that suggests the missing link might not have looked much like either a chimp or a human. *Ardipithecus*, seen by many as a human ancestor, lived about 4.4 million years ago. *Ardipithecus*'s anatomy suggests it spent a lot of its life in trees, unlike humans. But unlike chimps it lacked the features needed to swing below branches.

Ardipithecus (far right) revealed new information about the "missing link" – unlike the fake Piltdown Man fossils (above).

2/Shortcut: Darwin's vision of the natural world suggested that all living species – including humans – are related. Scientists gradually worked out that humans are most similar (genetically and anatomically) to chimpanzees. The earliest human-like fossils have been found in Africa. They are seven million years old, which suggests the human line split away from the chimp line just before this time. However, despite knowing where and when to look for the missing link – in African rocks that formed about seven million years ago – scientists will probably never find this animal. Ape fossils are very rare in such ancient rocks.

See also //

1 The theory of evolution by natural selection, p.6

14 The LUCA hypothesis, p.32

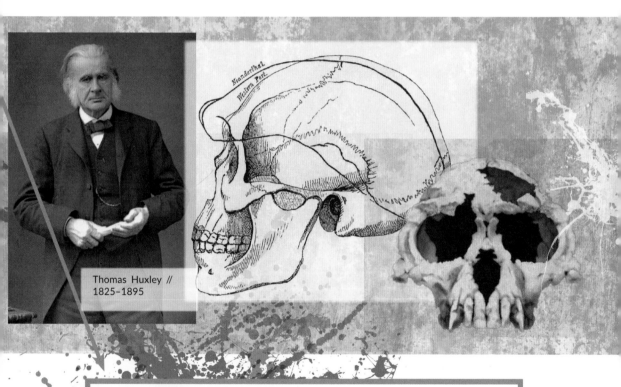

Thomas Huxley //
1825–1895

3/Hack: Evolutionary trees are like our personal family trees, but on a much grander scale. Biologists see humans and chimpanzees as siblings that were both "fathered" by the same ancestral species.

This ancestor is the missing link.

No.26
The drunken monkey hypothesis
Are there evolutionary benefits to boozing?

1/ Helicopter view: By the dawn of the 21st century, some biologists and medics had begun trying to understand human health from an evolutionary perspective. Addiction seemed to make sense when viewed this way too: our species has had little exposure to mind-altering drugs during its evolution, which might make us more prone to misusing them today. But **Robert Dudley** argues that one drug might break this rule.

Humans are primates, and as far as scientists can tell from the fossil record, fruit has been a component of the average primate diet for millions of years. This is the case even though fruit is often relatively difficult to find in the forest environments that primates usually inhabit. Even when a tree does come into fruit, those ripe and tasty treats quickly begin to rot.

This might be a key point, though, according to Dudley. When fruit rots and ferments, its sugars are converted into alcohol – which can then waft through the air. It might have been evolutionarily advantageous for primates to be attracted by the whiff of alcohol if that led them to trees where a few ripe fruit might still be found. In fact, argues Dudley, given that even ripe fruit contains some alcohol, it might have been beneficial for primates to learn to enjoy consuming the drug – potentially succumbing to its mind-bending effects in the process.

Dudley's **drunken monkey hypothesis** is not without its critics, but some genetic evidence suggests Dudley might be on the right track. In 2014, **Matthew Carrigan** and his colleagues discovered that a genetic mutation about 10 million years ago helped our distant ancestors break down the alcohol in their diet far more efficiently, which falls in line with the idea that boozing was beneficial even deep in prehistory.

Our love of alcohol might have deep prehistoric roots.

2/Shortcut: Millions of years ago, finding enough high-quality food was a significant challenge for our ancestors. Dudley argues that alcohol (both its scent and taste) might have become an important cue that alerted them to the possibility of a nutritious meal. As such, humans might be predisposed to consume as much alcohol as they can, because our bodies are still working on the assumption that alcohol-laced food is a rare treat. The problem is that today alcohol is widely available.

See also //

30 The obstetrical dilemma, p.64

3/Hack: The drunken monkey hypothesis takes a (pre)historical perspective on a modern problem. A love of alcohol is detrimental in today's societies,

but it might have been beneficial in the forests our ancestors occupied.

57

The savannah hypothesis
Why are we the upright ape?

1/ Helicopter view: Humans may be apes, but in many ways we look and behave very differently from chimpanzees and gorillas. Biologists have been trying for centuries to explain why our species began walking upright. For most of that time, savannahs have loomed large in their thinking.

In 1809, **Jean-Baptiste Lamarck** suggested that our distant ancestors lived in forests, before mysterious circumstances encouraged them to leave and adapt to life on open grasslands. **Charles Darwin** thought this was likely too. Sixty years after Lamarck, he speculated that it might have been environmental change that led to a reduction in tree cover and prompted a shift to savannahs. Darwin suggested that, with no trees to climb, our ancestors might well have opted to walk upright on two legs.

These ideas seemed to make so much intuitive sense – and came from such respected scientific figures – that many biologists felt it wasn't even necessary to formulate them as a scientific hypothesis. However, in 1960 **Alister Hardy** suggested an alternative narrative. Hardy thought it was significant that humans have little hair, possess a layer of fat beneath the skin, and have a natural ability to swim underwater – features usually seen in aquatic mammals. He suggested that a chunk of our evolution occurred in deep rivers or lakes rather than in forests or savannahs, and that some of our features are adaptations to a life in water. The idea became known as the **aquatic ape hypothesis**.

Hardy's idea has received little support from other scientists, but arguably it did help them to realize they needed to express the alternative "trees to savannah" model in formal terms. By the late 20th century the **savannah hypothesis** had emerged as the standard way to explain the origin of our unusual bipedal walking style.

Ancient footprints show that our ancestors have walked on two legs for millions of years, perhaps to adapt to savannah life.

2/Shortcut: Although the savannah hypothesis is popular, some scientists worry that it's more of a story than a testable idea. It suggests, for instance, that our ancestors began walking upright in order to get a better view of predators on the open plains. This, in turn, freed up their hands for new tasks, like fashioning tools, which might have led to the evolution of larger brains to make better use of those tools. But a lot of this is speculation. Worryingly for the hypothesis, fossil evidence discovered recently has begun to suggest that our ancestors began walking on two legs while still living in forest environments. In years to come, the savannah hypothesis might be rejected.

See also //
25 The missing link, p.54

3/Hack: The savannah hypothesis is a popular (but largely speculative) idea: it suggests our ancestors had to walk on two legs to survive on hostile grassy plains.

No.28
The grandmother hypothesis Celebrating the role of grannies

1/Helicopter view: In 1957, George Christopher Williams published a very influential paper on the physical deterioration that inevitably comes with age. Among other things, Williams considered the menopause. Most species continue reproducing until very late in life: humans are very unusual in that women enter a decades-long post-reproduction period usually beginning in the forties. Why is this so?

Williams argued that the menopause should be seen in context of the lengthy time it takes to raise children. At some point it makes more evolutionary sense for a woman to focus on caring for the children she has rather than trying to produce more.

In a 1989 book, **Kristen Hawkes** and her colleagues took Williams's work a step further. They had spent some time studying the Hadza, an indigenous group of people living in Tanzania, and they realised that post-menopausal women often worked harder to gather food than women of reproductive age.

Hawkes and her colleagues suggested post-menopausal women play a crucial role. The food they gather is shared with their adult children (particularly their daughters), who are too busy raising infants to gather enough food for their needs. If grandmothers improve the chances of their grandchildren surviving, they have a significant influence on population growth. And given that those grandchildren may carry the grandmother's genes for longevity, they too may live into old age and reinforce the grandmother role. The **grandmother hypothesis** was born.

The idea has become highly influential, but it is not without its critics. It is not clear when human lifespans extended to the point that grandmothers could remain active for many years beyond the menopause: some researchers think this is a relatively recent development, and so grandmothers cannot have been a key factor in infant survival throughout human evolution.

With "grandmothering" a woman can influence population growth ("r") even into her fifties .

2/Shortcut: Particularly in traditional societies, raising children is a long and difficult business for mothers: they must juggle childcare duties with the search for food. Hawkes and her colleagues think that, early in human evolution, mothers came to rely on their own mothers for much-needed help. The grandmothers who lived the longest and most active lives were most successful in this endeavour, and so their children and grandchildren (who carried their grandmother's genes for longevity) were most likely to survive. Gradually, humans evolved longer lifespans.

See also //

7 The concept of kin selection, p.18

22 The mitochondrial Eve hypothesis, p.48

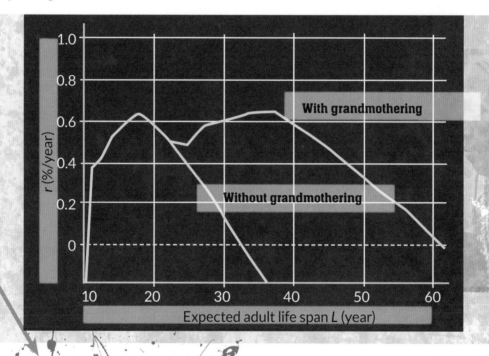

3/Hack: Human grandmothers are usually beyond reproductive age, but the grandmother hypothesis suggests they have been important throughout human evolution.

By foraging for food they might have increased the odds of their grandchildren thriving.

No.29

The cooking hypothesis
Did fire make us human?

Richard Wrangham // b. 1948

1/ Helicopter view: In 1999, **Richard Wrangham** came up with a novel and controversial explanation for why our early ancestors developed larger, more sophisticated brains. They did so, argued Wrangham, simply because they learned to cook.

The human evolutionary story is full of puzzles that have still to be answered. One of the most significant is why, several million years after our ancestors had adapted to life in relatively open grasslands, they suddenly began to look and behave differently. Early "ape-like" species called the australopiths walked on two legs and used tools like we do. But they also had strong chimpanzee-like arms, small chimp-like brains and large chimp-like teeth. About two million years ago the first "true" humans appeared and they were very different: they had shorter arms, larger brains and smaller teeth.

Wrangham thinks the trigger for this evolutionary event was the controlled use of fire – and particularly one important innovation it allowed: cooking. He argues that this early culinary revolution led to a dramatic change in the way human bodies worked and allowed brains to become larger.

Over the last 15 years or so, Wrangham and his colleagues have collected plenty of evidence in favour of this **cooking hypothesis** – but one significant problem remains. The idea predicts that scientists will find evidence of controlled fire at archaeological sites that date back two million years. So far, though, circumstantial evidence of controlled fire goes back only a million years, and strong evidence of controlled fire appears only 700,000 years ago.

The evolutionary trigger at the dawn of human life might have been the invention of cooking.

2/Shortcut: Chewing and digesting fibrous plants and raw meat is challenging. Cooking is essentially a way of predigesting food before it even touches our lips, so humans who cooked no longer had to grow large teeth and powerful jaws for heavy chewing. Even more important changes occurred in the gut. Our intestines are relatively short for an animal of our size, perhaps because our cooked food diet is more readily absorbed into the bloodstream. If early humans developed shorter guts, this would have freed up energy that could be redirected to growing other parts of the body – including the brain.

See also //

27 The savannah hypothesis, p.58

 3/Hack: Scientists and philosophers struggle to define exactly what it is that makes us human.

The surprising answer might be: cooking.

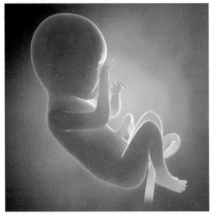

The obstetrical dilemma

Explaining the pain of childbirth

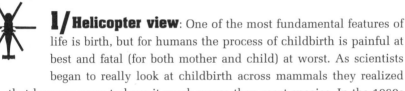

1/ Helicopter view: One of the most fundamental features of life is birth, but for humans the process of childbirth is painful at best and fatal (for both mother and child) at worst. As scientists began to really look at childbirth across mammals they realized that humans seem to have it much worse than most species. In the 1960s **Sherwood Washburn** suggested why.

Washburn looked at the features that define our species. Unlike other apes, he said, humans are most comfortable walking around upright on two legs. As our ancestors became more sophisticated toolmakers, said Washburn, they gained another uniquely human feature – an enormous brain.

Washburn suggested that these two features have essentially fought against each other during human evolution. The appearance of upright walking encouraged a re-organization of the human pelvis, which had the unfortunate consequence of constricting the birth canal. The appearance of large brains, meanwhile, required human foetuses to grow relatively large heads even before birth. Evolution was "selecting" both features – because both features allowed humans to thrive.

But the two features clash during childbirth: women evolved a narrow birth canal at the same time that their babies were evolving larger heads. Consequently, childbirth became a difficult and painful process early in human evolution and remains so to this day. Washburn described this as the **obstetrical dilemma**.

The idea is popular, although with a greater understanding of the process of childbirth some scientists have begun to question Washburn's elegantly simple hypothesis. For instance, emerging evidence suggests that women with wide hips that make childbirth a little easier are able to walk around on two legs just as easily and efficiently as women with relatively narrow hips.

2/Shortcut: Natural selection is like an invisible force pushing species to look and behave in a certain way — for instance, encouraging gazelles to grow longer, muscular legs so that the herbivores are more likely to outpace a predator. Washburn suggested that during human childbirth, natural selection effectively pushes in two opposing directions at once: our ancestors were more likely to survive not only if they were born with larger heads but also if they developed a narrow pelvis — hence the pain of childbirth.

The pain of human childbirth might have begun millions of years ago.

See also //

27 The savannah hypothesis, p.58

29 The cooking hypothesis, p.62

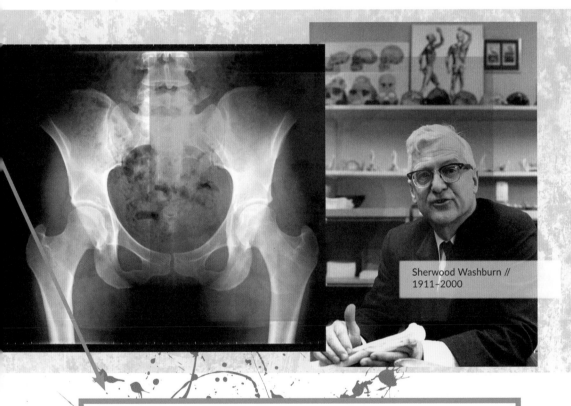

Sherwood Washburn //
1911–2000

3/Hack: The obstetrical dilemma provides an explanation for the difficulties of childbirth. It might seen odd that evolution would render childbirth occasionally fatal.

But enough children survived the process that human populations thrived.

No.31
The omnivore's dilemma
A rational explanation for irrational behaviour?

1/ Helicopter view: Parents of very young children sometimes face a daily battle to encourage their toddler to eat. In 1976, **Paul Rozin** provided an evolutionary explanation for this food fussiness.

Rozin argued that animals fall into two general dietary groups: specialists and generalists. A dietary specialist – a lion, for instance – has a very narrow sense of what a meal looks like. Over the generations, natural selection has favoured lions that know to recognize a gazelle as food.

Generalists, however, face a problem. They must eat a wide range of foods to meet all their nutritional requirements. They should have evolved an innate willingness to eat new plants or animals in their environment in order to meet their dietary needs. But those plants and animals might contain toxins – so omnivores should also have evolved an innate reluctance to eat new plants or animals. The omnivore's instincts are in opposition. Rozin called it the **omnivore's dilemma**.

Rozin realized that rats face this dilemma. From observations other researchers had made, Rozin suggested the rodents solve the problem by treating their bodies as a laboratory: a rat will nibble a tiny amount of a new food and then wait and see whether or not it brings on any ill-effects. If it leads to problems, the rat learns to permanently avoid that food – explaining why rats can be so difficult to poison.

Humans are omnivores too, so we face the same dilemma. By the late 1990s **Elizabeth Cashdan** had begun to suggest that it might be particularly acute in very young children, given that they have had little time to learn about what is and is not safe to eat. Fussy toddlers might simply be obeying their instincts.

Lions know exactly what to eat – but humans and other omnivores have almost limitless options.

2/Shortcut: Toddlers have just learned to walk and gain some independence from their parents, giving them freedom to explore their environment alone. But they haven't gained enough experience to work out which plants and animals in that environment are safe to eat and which are toxic. According to Cashdan, it might have been beneficial for toddlers to avoid trying new foods, just in case they were harmful. That reluctance might still exist in toddlers today.

See also //

26 The drunken monkey hypothesis, p.56

3/Hack: The omnivore's dilemma suggests we pay a price for our broad diet: it has left us wary of trying new and potentially toxic foods.

Toddlers are particularly sensitive to the risks.

No.32
The out of Africa hypothesis An evolving understanding of humanity's origins

 1/Helicopter view: After **Charles Darwin** formulated his **theory of evolution by natural selection**, biologists gradually came to accept that humanity must have an evolutionary past. In the 150 years since Darwin wrote down his theory, an abundance of fossil evidence has come to light, helping to build a picture of the human evolutionary tree and the origin of our species.

The fossil record suggests that the earliest chapters of the human evolution story took place in Africa. The oldest, and most ape-like, of our ancestors seem to have lived in East and Central Africa at least five million years ago.Between two and three million years ago these ape-like species evolved into species that were recognizably human. Shortly after, these early humans began spilling out of Africa and spread across Eurasia.

Then comes the controversy. According to a model advocated by **Milford Wolpoff** among others, these human populations remained in contact, interbreeding with each other such that our species – *Homo sapiens* – essentially appeared across Africa and Eurasia at the same time. This is the **multiregional hypothesis**.

The alternative view, championed by scientists including **Chris Stringer**, is that all living humans belong to a tightly defined group that first evolved a few hundred thousand years ago in Africa, and then spread across the world, largely replacing the ancient humans that occupied other areas. This is the **out of Africa hypothesis**.

With the advent of cheaper genetic sequencing technology, scientists can now afford to study the genes of people all across the world. These studies seem to confirm that all living people are very closely related to one another – and that they all share an ancestor within the last few hundred thousand years. Consequently, most researchers now favour the **out of Africa hypothesis**.

Our species evolved in Africa about three hundred thousand years ago and then spread across the world.

2/Shortcut: There are two ways of thinking about the human evolutionary tree. The first is to lump everything that looks recognizably human into one grand species that appeared two million years ago and gradually morphed into the modern form that populates the world today. The second is to recognize that living humans are anatomically and behaviourally distinct from ancient humans like the Neanderthals, and to put them in a species that evolved relatively recently in prehistory. Today, most scientists prefer this second option.

See also //

22 The mitochondrial Eve hypothesis, p.48

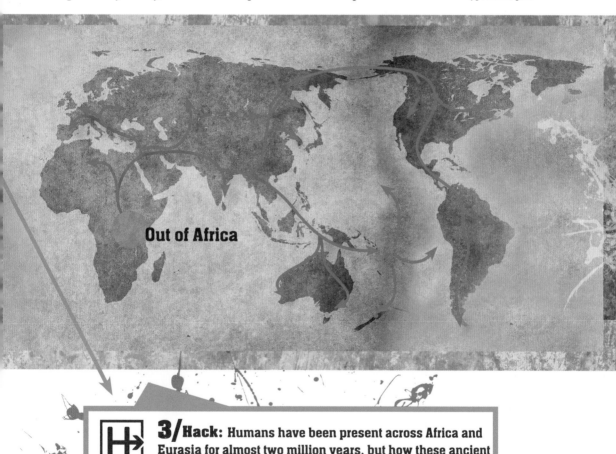

Out of Africa

3/Hack: Humans have been present across Africa and Eurasia for almost two million years, but how these ancient humans relate to living people was once unclear.

The out of Africa hypothesis argues living people stem from a group that appeared in Africa just a few hundred thousand years ago.

No.33
The father tongue
hypothesis How languages spread
around the world

 1/Helicopter view: In 1997, **Laurent Excoffier** and **Estella Poloni** discovered that human DNA might provide a new insight into the way languages have spread around the world.

Many scientists now think our species – *Homo sapiens* – originated in Africa in the relatively recent past and began migrating to other world regions 100,000 years ago. This suggests any differences – such as language – between people now living in different world regions must have begun to appear during those migrations.

Linguists have established a "family tree" that incorporates most languages. For instance, English and German fall on the Germanic branch, and have more distant ties to Spanish and French, on the Romantic branch. But all four languages are similar at a deep level: they lie within a vast Indo-European language family that includes most modern European languages as well as several spoken in Asia.

Excoffier and Poloni asked whether genes offered a new way to understand how languages branched from one another. They sequenced DNA from the Y chromosome, a

small chunk of genetic material found only in males that passes exclusively from fathers to sons. A global map of Y chromosome sequences can help reveal how men (but not necessarily women) migrated in prehistory. Another type of DNA in human cells – mitochondrial DNA – passes from mothers to their children, and can give a sense of the prehistoric migrations of women.

Perhaps unsurprisingly, the "ancient male" and "ancient female" migration maps are fundamentally similar. But there are differences – and Excoffier and Poloni realized the male migration map was better able to explain the distribution of languages in Europe and parts of Africa. Put simply, it seems that people tended to adopt the language of their father rather than of their mother. The idea is known as the **father tongue hypothesis**.

Languages often appear to pass down the paternal line.

2/Shortcut: DNA suggests prehistoric men and women didn't spread across the world in quite the same way. Sometimes, for instance, groups of men went marauding. They ended up settling in new areas and taking local women as wives. Excoffier and Poloni found the subtle echoes of these episodes in modern DNA: signs that migrating or invading groups of men often "recruited" local women, who then adopted the invaders' culture and language. The children from these unions took their (invading) father's culture too and helped it become established in the area. They adopted their father's tongue.

See also //

22 The mitochondrial Eve hypothesis, p.48

32 The out of Africa hypothesis, p.68

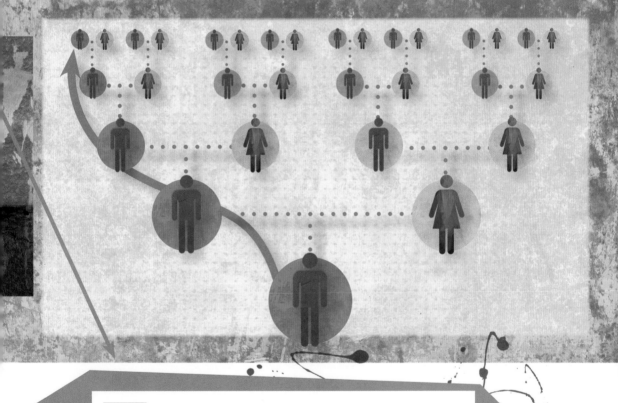

3/Hack: The father tongue hypothesis challenges our understanding of language spread.

It suggests that the European family of languages was brought to the continent primarily by men.

No.34
The black swan problem
Will tomorrow really come?

1/ Helicopter view: Reportedly, the citizens of 16th century London were fond of describing impossible events as "black swans". The phrase seemed completely justified: Europe was full of swans and all of them were white. Then, in 1697, Dutch explorers sailing up a river in Western Australia made an astonishing discovery: a black swan.

In the annals of scientific investigations, the discovery of the black swan is a mere footnote. But in the 20th century it became famous for the spotlight it shone on a problem at the heart of all scientific endeavour: the problem of induction.

Philosophers have been aware of the problem of induction for centuries, and the threat it poses to science. In essence, it describes the way we use past experiences to justify predictions about the future: the Sun always rises in the morning, so we assume it will rise again tomorrow. Scientists rely on induction to form general laws or theories. But is that a valid strategy?

In the 18th century, **David Hume** concluded that it was. He accepted that we cannot actually prove that our generalizations about the world are correct – that snow is always cold, for instance – but for some mysterious reason our instincts encourage us to accept these generalizations. And that's enough to justify the continued use of inductive reasoning.

Karl Popper // 1902-1994

In the 1950s **Karl Popper** rejected this conclusion. He argued we cannot ignore the problem of induction. His solution was to suggest that scientific research is really about formulating theories that can be disproved – an idea that implies science does not and cannot establish fundamental truths. He cited the **black swan problem** as one way to distinguish between what science is and what it is not. However, although Popper's work was very influential, it was not universally accepted. Arguably, the problem of induction is still a significant one.

Black swans symbolise a problem at the core of science.

72

2/Shortcut: According to Popper, scientists should concern themselves with generating theories that can be disproved, not with trying to arrive at universal truths. A non-scientific response to the observation that European swans are white is to conclude: A non-white swan is impossible. The scientific response would be to formulate a testable (and falsifiable) theory: I hypothesize that all swans found in future will be white. As such, said Popper, scientific theories remain valid until they are disproved – but strictly speaking they can never be proved.

See also //

37 The concept of uniformitarianism, p.78

100 The panspermia hypothesis, p.204

3/Hack: Popper's black swan problem is about the fundamental nature of science.

Arguably, scientific research cannot prove a theory is true, it can only confirm it is false.

No.35
The concept of deep time
When our planet began to show its age

1/Helicopter view: One spring morning in 1788, **James Hutton** and **John Playfair** took a day trip from Edinburgh to look at the geology of the Berwickshire coast. Near a site called Siccar Point, they found what they were looking for – an arrangement of rocks that helped Hutton demonstrate that Earth was far, far older than most had imagined possible.

The prevailing, religious-inspired view in 18th-century Europe was that our planet was no more than a few thousand years old. Although some geologists had begun to question that age, Hutton's thoughts on the subject would ultimately be the most influential. This is largely because Hutton was among the first to arrive at a recognizably modern understanding of the way different types of rock actually form. He theorized that the Earth is constantly making and destroying continents. Erosive forces destroy rocks and return them as sediment to the sea. This sediment builds up in horizontal layers on the seafloor and turns to rock – a process that could, in principle, take many thousands of years. Eventually the sedimentary rock pile is so thick and so heavy that the lower layers heat and melt under pressure, forming buoyant molten lava that forces the overlying rock up above the waves to form a new continent.

This isn't exactly the rock cycle as geologists recognize it today, but it is a cyclical process – one that Hutton said could continue essentially indefinitely. There was, in Hutton's words "no vestige of a beginning – no prospect of an end".

Hutton presented his theory to Edinburgh's leading scientific society in 1785. His careful geological observations in the years that followed – including his trip to Siccar Point – provided empirical evidence to back his claims. Even so, when Hutton died in 1797 his ideas were not widely known. Playfair redrafted and revised Hutton's ideas in 1802 and their influence grew. The **concept of deep time** was on its way to broad acceptance.

In some cliff faces, horizontal layers of rock sit on top of vertical layers.

2/Shortcut: At Siccar Point, there is a series of sandstone layers that seem to have been tilted slightly. They rest on top of another series of sandstones – these ones almost vertical. Hutton reasoned that sandstones originally form as horizontal layers in water – a process that might take thousands of years. He realized the geology at Siccar Point must show two different episodes of sandstone formation. After the first sandstones had formed they were violently rotated (to near-vertical) and forced upwards to create dry land. The surface of this land then wore smooth, before being submerged again – allowing for a second episode of horizontal sandstone formation. Finally, the whole sequence was forced above the waves again and tilted slightly to create the modern landscape.

See also //

37 The concept of
uniformitarianism,
p.78

83 The concept of
radiometric dating,
p.170

James Hutton // 1726–1797

3/Hack: Early scientists assumed, on religious grounds, that our planet is very young.

Hutton's work helped convince them that the Earth is extraordinarily old.

No.36
The theory of catastrophism
When the world changes overnight

1/ Helicopter view: The geologists of the early 19th century knew that an awful lot had occurred in our planet's past – but the general assumption was that the Earth was just a few thousand years old. How could so much geological history be squeezed into so short an interval of time?

European science at this time was still governed by Biblical assumptions – among them the idea that the Old Testament was a literal history of the world, and that this history stretched back roughly 6,000 years. But some of the evidence geologists were uncovering in the fossil record sat uncomfortably with religious beliefs. In the late 18th century, for example, **Georges Cuvier** began to argue that species could become extinct – something that religious teachers assumed to be impossible.

Cuvier's reading of the fossil record suggested there had been several extinctions. What's more, extinction events seemed to be catastrophic, affecting not just one species but entire ecosystems. He argued that Earth periodically experiences a rapid and dramatic ecological event that wipes out most species in an instant, leaving the world more or less empty for colonization by the few survivors. The theory provided a convenient way to compress the eventful history of life on Earth into a few thousand or a few million years. **William Whewell** gave Cuvier's theory its name: **catastrophism**.

Despite initial popularity, catastrophism fell out of favour later in the 19th century. In fact, it languished for most of the 20th century too. Its fortunes changed in the 1980s, when **Luis** and **Walter Alvarez** uncovered evidence that a calamitous asteroid impact occurred 66 million years ago, coincident with a mass extinction that destroyed entire ecosystems, and killed all of the large dinosaurs. Their work led to a renewed interest in catastrophism.

2/Shortcut: Cuvier spent time studying the geology of the area around Paris with **Alexandre Brongniart**. Their research was some of the earliest to establish the science of stratigraphy – the idea that, generally speaking, older layers of rock are found at the bottom of a sequence and younger rocks above. Cuvier and Brongniart noticed that one layer of rock often contained completely different fossils from the layer it lay above, and from the layer it rested beneath. The ecological changes were abrupt. Cuvier thought this betrayed the occurrence of short, sharp ecological catastrophes.

An asteroid impact can lead to an instant global catastrophe.

See also //

37 The concept of uniformitarianism, p.78

Georges Cuvier // 1769–1832

3/Hack: Most geologists think Earth's biosphere generally takes millions of years to change in any appreciable way.

But catastrophism suggests that occasionally the world changes dramatically overnight.

The concept of uniformitarianism
Unlocking the past

Sir Charles Lyell // 1797–1875

1/ Helicopter view: In the 1830s, **Charles Lyell** revolutionized the study of the ancient Earth. To his supporters, including **Charles Darwin**, Lyell had taken geology, a subject steeped in religious dogma, and moulded it into a rational scientific discipline.

Lyell had been taught by **William Buckland**, a champion of the idea that Earth's past was occasionally extremely violent. But Lyell became disillusioned with the idea, particularly when Buckland began arguing that the biblical flood caused the most recent catastrophe in Earth's history. Lyell thought geology should not appeal to supernatural forces.

He was heavily influenced by the geological ideas set out by **James Hutton** in the 1780s, which suggested the geological record showed evidence of a (potentially) never-ending cycle of geological processes. Hutton's ideas seemed to be arguing for the uniformity of forces acting through time – an idea that was in line with **Isaac Newton**'s successful theories on the uniformity of forces across space.

Lyell liked this idea. He also liked the philosophical works of **David Hume** – particularly the idea that scientists should be encouraged to reason inductively, or infer general rules from a limited set of real observations. Combining aspects of Hume's and Hutton's work, Lyell devised a new way to think about the geological and fossil records. Put simply, he argued that the slow but steady processes we see operating today are all that can or should be used to account for the features of the geological record. Lyell's idea is often summarized using the phrase "the present is the key to the past". **William Whewell** – who was actually highly critical of Lyell's concept – gave it a name: **uniformitarianism**.

Uniformitarianism argues that the geological processes occurring today can explain anything in Earth's past.

2/Shortcut: Lyell was a firm believer that Earth's prehistory is cyclical in nature – that features of the landscape gradually form and then gradually disappear again. He said evidence of this cyclical process of creation and destruction exists around us, if we look carefully. For instance, Lyell knew of the existence of earthquakes, and how their movement can often remould landscapes, thrusting land into the air to create small cliffs and hills. He also knew that erosion can gradually wear down those cliffs and hills until they are virtually indiscernible. Lyell argued that processes like these, occurring over an unfathomably long timescale, could account for anything in the geological record.

See also //

34 The black swan problem, p.72

35 The concept of deep time, p.74

36 The theory of catastrophism, p.76

56 Newton's law of universal gravitation, p.116

 3/Hack: According to Lyell, all the tumultuous events documented in the geological record can be explained using the slow geological processes occurring on Earth today:

to understand Earth's past, just look around you.

No.38

The dynamo theory

Unravelling the mysteries of Earth's core

Joseph Larmor // 1857–1942

1/Helicopter view: For centuries, navigators have relied on Earth's magnetic properties – but not until scientists began to explore Earth's interior did it become clear exactly why our planet is magnetic.

As far back as 1600, **William Gilbert** was aware that some of Earth's minerals are naturally magnetic. He concluded that Earth must be, in effect, an enormous permanent magnet. Later in the 17th century, scientists began to suspect things were more complicated, not least because their careful measurements showed that the magnetic field varied slightly from day to day – something that shouldn't happen to the field around a permanent magnet. What's more, it was becoming clear that Earth's interior is hot, and heat destroys permanent magnets.

Things finally became a little clearer a century ago. By analysing the way that seismic waves from earthquakes travel through the Earth, **Richard Oldham** concluded, in 1906, that the outer portion of Earth's core must be liquid – almost certainly molten iron. In 1936, **Inge Lehmann** concluded the inner core is solid using broadly similar techniques.

By that time, **Joseph Larmor** had already suggested how the core could generate Earth's magnetic field. Heat from the inner core could create convection currents in the liquid outer core. This churning molten iron would naturally generate electrical currents. Given the link between electricity and magnetism the electrical current would generate a magnetic field, which could encourage further movement of the (magnetic) liquid iron, generating its own magnetic field, and so on. The feedback loop might ultimately help form one very powerful magnetic field.

Walter Elsasser and **Edward Bullard** further developed the idea in the 1940s and 1950s. Their work helped ensure that the **dynamo theory** became a mainstream idea.

The dynamo theory links Earth's magnetic field to our planet's iron-rich core.

2/Shortcut: Earth's core behaves a little like a spherical lava lamp. A solid and very hot inner core helps heat the liquid outer core, creating convection currents in the liquid. However, this is a lava lamp with a difference: the liquid being moved is iron-rich and electrically conductive – as it moves it generates electric currents, which in turn generate magnetic fields. As a consequence of Earth's rotation, the currents (and their associated magnetic fields) become aligned and combine: this combined field is so strong we feel its effects at Earth's surface, even though the Earth's core is 3,000 kilometres (1,864 miles) below our feet.

See also //
60 Maxwell's equations, p.124

3/Hack: The dynamo theory suggests heat from the inner core causes molten iron in the outer core to flow.

This ultimately generates Earth's powerful magnetic field.

No.39
The theory of continental drift

The world begins to move

Alfred Wegener // 1880–1930

1/ Helicopter view: As long ago as the 16th century, map-makers had noticed the similarity between the eastern coastline of the Americas and the western coastline of Europe and Africa. **Abraham Ortelius** speculated that the landmasses had once been connected, but had then been torn apart through the action of earthquakes or floods. In the 20th century this idea was renewed, revised – and eventually accepted. That scientific journey began with **Alfred Wegener**.

Wegener was a meteorologist and polar explorer rather than a pure geologist. But he was well-read, and he was aware that other scientists were puzzling over the geological similarities between the Old World and the New World – particularly the fact that very similar fossils of land-living animals and trees were turning up in West Africa and in Brazil. He was also aware of the leading theory to explain this puzzle: that the landmasses had once been linked by large land bridges, but that these had sunk beneath the waves of the Atlantic Ocean long ago.

Wegener read about ongoing research into the density of rock types, and realized the land bridge theory was wrong. He argued in 1912 that the continents were mobile, as Ortelius (and several others down the years) had suggested.

The idea certainly did seem to solve many geological mysteries. Nevertheless, geologists dismissed Wegener's idea. Crucially, in his 1915 book *The Origin of Continents and Oceans*, Wegener failed to provide a convincing mechanism to explain how vast continents could move. Today, Wegener is remembered as a visionary who helped propel the scientific community towards making one of the most significant breakthroughs in the history of geology. But when he died in 1930, few geologists were yet prepared to accept Wegener's **theory of continental drift.**

2/Shortcut: In the early 20th century many geologists thought that the strikingly similar fossil record of West Africa and South America could be explained by a land bridge that once linked the two and that allowed animals and plants to migrate between the continents. By 1912, Wegener knew this idea couldn't work. Rock beneath the oceans is much denser than the rock that forms continents: it didn't make sense for a vast, buoyant (continental) land bridge to sink permanently to the bottom of the ocean. This density contrast was also at the mechanical heart of Wegener's theory. He imagined continents as giant low-density rafts that could plough through the denser ocean crust – an idea most other geologists considered absurd.

The east coast of the Americas looks similar to the west coast of Europe and Africa.

See also //

40 The theory of seafloor spreading, p.84

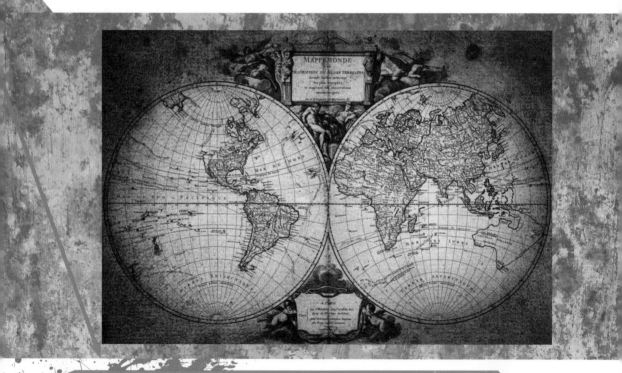

 3/Hack: Continental drift argues that Earth's vast continents move around very slowly.

But the theory did not offer a convincing explanation for how continents drift.

No.40
The theory of seafloor spreading
The ocean floor reveals its secrets

1/Helicopter view: Early in the 20th century, most geologists found it difficult to accept that Earth's giant continents could move across the surface of the planet. The research that helped changed their minds came from an unexpected place: studies of the ocean floor.

Many scientists at this time worked on the assumption that the seafloor was flat, featureless and not particularly interesting – even though scientific voyages in the 19th century had hinted at a more varied picture. In the 1950s it became clear just how variable the ocean floor really is. Using data from US navy surveys, **Marie Tharp** and **Bruce Heezen** began to produce the first maps of the ocean floor.

The work revealed that very long submarine mountain chains stretched roughly down the centre of oceans including the Atlantic and Indian. They became known as mid-ocean ridges. Even more curiously, running right down the centre of each mid-ocean ridge was a deep valley: a rift.

Heezen and other scientists working on ocean geology – in particular **Harry Hess** – became convinced that magma from deep inside the Earth constantly wells up to the surface and cools to form new rocky oceanic crust at the mid-ocean ridge rifts. Because oceanic crust is being created constantly, oceans grow: they become wider. This became known as the **theory of seafloor spreading**.

Hess realized that this process could provide a convincing mechanism to explain continental drift. **Alfred Wegener** had suggested that continents move by actively ploughing through the oceans. Hess suggested an alternative. He argued that the continents are passive: all of the active dynamism that Wegener recognized in his **theory of continental drift** actually occurs at the bottom of the ocean. To put it another way, the continents drift because the seafloor spreads.

Mid-ocean ridges zigzag down the centre of oceans, including the Atlantic.

2/Shortcut: It might help to think of mid-ocean ridges as gigantic 3D printers. A mid-ocean ridge trending north–south is constantly "printing" new crust both to the east and to the west. The rift running along the length of the ridge is a reservoir of 3D printer ink – hot magma. It cools to form new crust. Half gets attached to the slab of crust to the east of the ridge, half to the slab on the west. Tectonic forces gradually pull those two slabs apart, allowing more magma to well up at the rift and cool into more crust, and so on.

See also //

39 The theory of continental drift, p.82

Marie Tharp // 1920–2006

3/Hack: Seafloor spreading suggests oceans can, very slowly, grow wider. In doing so they can push the continents on either side of the ocean further apart.

Seafloor spreading offers an explanation for continental drift.

No.41
The theory of plate tectonics
Geology's grand unifying theory

 1/Helicopter view: A scientific revolution that had begun in the 1910s with **Alfred Wegener** reached its tipping point in the middle of the 20th century: geologists were poised to accept that the Earth's surface is in constant motion.

Wegener had insisted that continents move, although he failed to come up with a convincing mechanism to explain how they did so. By the 1950s, geologists had begun to suspect the answer lay with mid-ocean ridges –

submerged mountain chains that run down the centre of most oceans. But more evidence was needed. It soon arrived.

In response to the 1963 Partial Test Ban Treaty, authorities set up a global network of earthquake monitors to keep tabs on nuclear weapons tests. The network also helped geologists produce a global map of natural earthquakes. This revealed that quakes aren't distributed randomly: they trace out long lines, some of which coincide with mid-ocean ridges, others with deep trenches known to exist in the ocean. It looked as if Earth

had a hidden geography. There was a series of "plates" that seemed to jostle at their margins, generating intense earthquakes.

At roughly the same time, **Keith Runcorn** and others were revealing another piece of evidence. Magma often contains magnetic minerals, and as the magma cools into rock the minerals become fixed in alignment with Earth's magnetic field. Looking at this alignment in rocks of different ages showed something puzzling: whole continents seemed to have moved or even rotated in the past – just as Wegener had argued decades earlier. By the late 1960s few geologists denied this growing body of evidence. The **theory of plate tectonics** had become accepted science.

Nuclear weapons tests inadvertently led to an understanding of Earth's tectonic plates.

2/Shortcut: Earth's crust is like a complicated jigsaw puzzle, currently comprising seven large pieces (plates) and several much smaller ones. Over time, some of those pieces grow larger as new crust forms at mid-ocean ridges. As they grow, other pieces must shrink to make room. This can happen where two plates meet: one might be pushed below the other and sink back into Earth's interior, a process that slowly destroys the plate. The continents – which are simply thicker parts of the plates – are inadvertently pushed, pulled and rotated as the plates jostle against each other. Sometimes, continents collide to create high mountains. At other times, continents are torn apart along great rift valleys that will eventually turn into new oceans.

See also //

38 The dynamo theory, p.80

39 The theory of continental drift, p.82

40 The theory of seafloor spreading, p.84

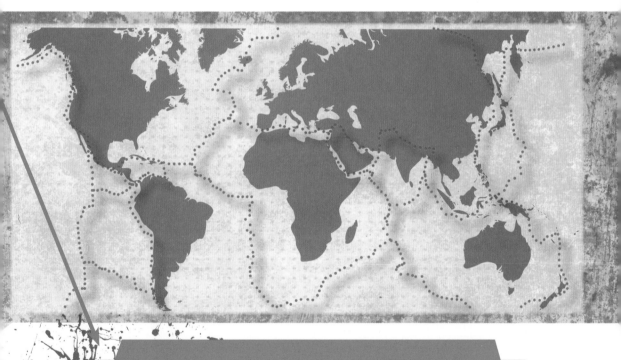

3/Hack: Almost everything geologists observe – from tall mountains to deep oceans to the distribution of fossils – makes sense through the prism of plate tectonics.

It is the grand unifying theory of geology.

No.42
The Milanković theory
Natural climate change explained

1/Helicopter view: European scientists of the early 18th century had many curious problems to grapple with. One was explaining the existence of giant boulders scattered across the landscape, particularly in Alpine areas.

In the 1740s, **Pierre Martel** journeyed to one Alpine valley and asked the locals for their thoughts. He was told that the valley's glacier had once been much larger. The giant boulders had formerly been locked in ice, which must have been responsible for moving them. Martel was one of the first scientific researchers to recognize the possibility that our planet had once experienced ice ages.

By the 20th century geologists had established that there have been several ice ages. The most recent ice age became the best known. It began about 2.5 million years ago, and it seemed to be characterized by several colder intervals (glacials) separated by warmer periods (interglacials). Why this complex pattern?

Milutin Milanković (sometimes written **Milankovitch**) set out to solve this mystery. In the 1920s, he began formulating a theory that suggested subtle but regular variations in the Earth's movement as it orbits the Sun lead to small changes in the way solar energy hits the planet. These differences, he argued, should be enough to trigger the shifts from glacials to interglacials.

Milutin Milanković // 1879–1958

Milanković's calculations were detailed, but for 50 years no one knew whether they were meaningful. In the mid-1970s, three scientists – **James Hays**, **Nicholas Shackleton** and **John Imbrie** – decided to find out. They studied mud that had been accumulating at the bottom of the Indian Ocean, using its chemistry and the tiny microfossils it contained to work out how Earth's climate has varied over the last 450,000 years. The pattern of variation was strikingly similar to the pattern Milanković had predicted on theoretical grounds. Many geologists became convinced by his theory.

The three ways in which Earth's movement changes.

2/Shortcut: Milanković knew that the Earth's movements vary slightly but regularly on three different time scales: a 21,000-year cycle, a 41,000-year cycle and a 96,000-year cycle. Combined, the three cycles predictably alter the distance between the Earth and the Sun, and the Earth's exact orientation relative to the Sun. Milanković calculated that at certain points in these cycles, our planet is positioned and angled in such a way that the amount of incoming solar radiation hitting a latitude just below the Arctic circle is too weak even in summer to melt ice forming there. This means that more and more ice can accumulate each winter, triggering a glacial episode. At other points in the cycle, more solar radiation hits this latitude and melts ice, leading to an interglacial.

See also //

57 Newton's laws of motion, p.118

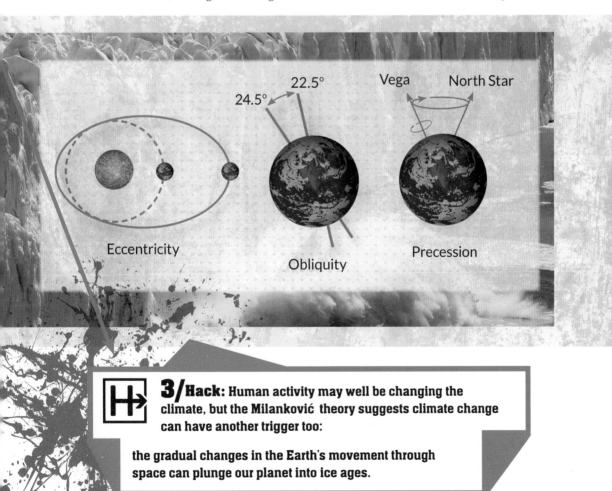

24.5°   22.5°   Vega   North Star

Eccentricity   Obliquity   Precession

3/Hack: Human activity may well be changing the climate, but the Milanković theory suggests climate change can have another trigger too:

the gradual changes in the Earth's movement through space can plunge our planet into ice ages.

No.43
The snowball Earth
hypothesis When the tropics froze

1/Helicopter view: **Douglas Mawson** knew how to recognize evidence of former glacial activity in the rock record. Naturally, then, when Mawson found such evidence in ancient rocks from South Australia he concluded that the area had once been blanketed in ice. There was just one problem with this idea: South Australia's relative proximity to the equator. In the 1940s, Mawson suggested there had once been an intense ice age that affected even tropical regions.

Mawson's peers dismissed his idea, particularly given mounting evidence that continents drift. It was easy for Mawson's critics to argue that his glacial deposits dated to a time when Australia had been nearer the poles, where everyone accepts that ice can form.

But in the 1960s, after Mawson's death, the study of plate tectonics would provide key evidence in favour of his global ice age idea. Geologists began to use magnetic minerals in rocks of different ages to work out where continents had been located at various points in the past. In 1964, **Brian Harland** established that glacial deposits

had, after all, formed on land near the tropics about 650 million years ago – the age of Mawson's Australian rocks.

It still wasn't clear how a global glaciation could occur – or, just as importantly, how it would ever end. But in 1992, **Joseph Kirschvink** proposed a mechanism, and made predictions about what sort of geological evidence would be left as a "snowball Earth" – his term for global glaciations – formed and then melted. An influential study published in 1998 by **Paul Hoffman** and his colleagues found evidence for several of Kirschvink's predictions. Many geologists now accept the **snowball Earth hypothesis**.

Some geologists think Earth can sometimes turn into a giant snowball.

2/Shortcut: Kirschvink argued that it was probably necessary for all the continents to be clustered at or near the equator to form a snowball Earth. This would cool the planet because land is relatively pale and so reflects a lot of solar energy back into space. However, once a snowball Earth forms Kirschvink said it would help bring about its own destruction. Plate tectonics would continue to operate, meaning volcanoes would spew carbon dioxide into the atmosphere. But on snowball Earth, this carbon dioxide wouldn't be mopped up by photosynthetic organisms, most of which would have died as the ice formed. As carbon dioxide levels rose, a powerful greenhouse effect would have kicked in. The snowball Earth would have melted.

See also //

40 The theory of seafloor spreading, p.84

41 The theory of plate tectonics, p.86

SIR DOUGLAS MAWSON   3754·2

Sir Douglas Mawson // 1882–1958

3/Hack: The snowball Earth hypothesis argues that a rare set of natural circumstances can trigger global cooling on an extraordinary scale.

During these intervals even the tropics can freeze.

No.44
The Gaia hypothesis

A planet built for (and by) life

James Ephraim Lovelock // b. 1919

1/Helicopter view: In 1965, while working at NASA, **James Lovelock** had a chance to view some of the first data on the composition of the Martian atmosphere. The results revealed a clear difference between Earth and its two neighbours, Venus and Mars. While Earth's atmosphere contains about 21 per cent oxygen and 0.04 per cent carbon dioxide, both Mars and Venus appear to have atmospheres dominated by carbon dioxide. Lovelock realized the results prompted a question: why is Earth's atmosphere so different?

Theoretically speaking, oxygen should not build up in a planet's atmosphere: it should react chemically with other gases and vanish. Lovelock knew that oxygen is abundant in Earth's atmosphere because it is constantly being generated by living things: Earth's biosphere has had a planet-wide impact on the atmosphere. It suddenly dawned on him that living organisms are, effectively, regulating the composition of the atmosphere.

The more he thought about the subject, the more Lovelock became convinced that life influences our planet, keeping the atmosphere in a relatively stable state that is conducive to the continued survival of life on Earth. It's an idea that might actually stretch back at least as far as the late 18th century: **James Hutton** was convinced of a fundamental connection between geological and biological processes, for instance.

Lovelock lived in the same English village as the novelist **William Golding**, who suggested a name for the idea: the **Gaia hypothesis**. Lovelock's idea was controversial in the 1960s and it remains so to this day. For one thing, it has become clear that there are intervals in prehistory when life spectacularly failed to keep Earth's conditions in a stable state conducive for life. Lovelock's idea has even triggered a counter-hypothesis. Nevertheless, the Gaia hypothesis remains popular outside the scientific community.

Has life on Earth coevolved with our planet?

2/Shortcut: One of the key mysteries Lovelock tackled through his Gaia hypothesis is related to the "faint young Sun paradox". Early in Earth's history the Sun burned less intensely than it does now, but Earth's early geology shows that water flowed on our planet's surface, suggesting surface temperatures were broadly similar to today. Many scientists explain this paradox by arguing that Earth's early atmosphere was rich in greenhouse gases like carbon dioxide that helped trap heat. However, life tends to feed on these gases, which could have weakened the early greenhouse effect and cooled the planet. Lovelock argued that life must have found a way to regulate the atmosphere so that Earth's temperature has remained stable and "life friendly" for billions of years.

See also //

43 The snowball Earth hypothesis, p.90
45 The Medea hypothesis, p.94

3/Hack: The Gaia hypothesis argues that life has coevolved with our planet.

According to the idea's advocates, life has kept Earth's physical conditions ideally suited for habitation.

No.45
The Medea hypothesis
An antidote to Gaia

1/Helicopter view: By the early 2000s the **Gaia hypothesis** (page 92) had risen almost to the level of religion in some circles. Although scientists continued to debate its merits, evidence was beginning to cast serious doubts over the idea. In 2009, **Peter Ward** suggested this evidence didn't simply undermine the Gaia hypothesis – it actually suggested Earth was "anti-Gaian".

An initial strength of the Gaia hypothesis was that it seemed to explain why Earth has remained stable and habitable for billions of years. Discoveries in the final few decades of the 20th century were compatible with the idea. For instance, in the 1980s geologists found evidence that the famous mass extinction that killed the large dinosaurs may have been triggered by an asteroid impact. Scientists began to argue that all mass extinctions were triggered by asteroids. This idea fit well with the Gaia hypothesis because it suggested that mass extinctions were essentially random "acts of God" – Earth (and Gaia) was in no way responsible for them.

But by the dawn of the 21st century opinions were shifting. Geologists abandoned the idea that mass extinctions are always triggered by asteroids: they simply failed to find evidence to back the hypothesis.

Scientists began to argue that mass extinctions must usually be triggered by events on Earth itself. This strongly suggested to Ward that life might have a role to play in triggering ecological catastrophes. The Gaia hypothesis is based on the idea that biological species help keep Earth habitable. Ward argued the opposite: that biological species might actually destabilize Earth and make it inhospitable, triggering mass extinctions. He called this idea the **Medea hypothesis** – Medea being a character from Greek mythology famous for killing her children.

The Medea hypothesis argues that life will sometimes turn on itself.

2/Shortcut: Many geologists now think that on more than one occasion, Earth became so cold that the planet essentially froze over, probably causing an extinction. Ward points out that each proposed Snowball Earth event occurred a few hundred million years after the appearance of new forms of photosynthetic life that could suck carbon dioxide out of the atmosphere more efficiently. Carbon dioxide is a powerful greenhouse gas: when its atmospheric level drops, the planet cools down and ice sheets form. In other words, Ward argues that life destabilized the environment, and triggered a global catastrophe.

See also //
36 The theory of catastrophism, p.76
43 The snowball Earth hypothesis, p.90

3/Hack: According to the Medea hypothesis, species on Earth are far more likely to be wiped out through the action of other organisms than through asteroid impacts or other natural disasters.

In other words, danger comes from within.

No.46

Robert Paine // 1933–2016

The keystone species concept
Not all species are ecologically equal

1/ Helicopter view:

Ecologists working in the 1960s knew that natural ecosystems can be incredibly complex: dozens of species might interact with each other in all sorts of different ways. The assumption was that this complexity leads to stability. In 1969, **Robert Paine** questioned that assumption. He argued that even complex ecosystems could change dramatically by the removal of just one species.

It's easy to see why ecologists assumed complex ecosystems are stable. An ecological food web is a network, a little like the internet. If one internet cable is destroyed, information can usually find another path to move from computer A to computer B. Likewise, if one species in the ecosystem disappears, energy should still be able to flow through the food web more or less as it did before. The ecosystem as a whole shouldn't really change.

Paine's ecological work suggested this isn't necessarily the case. He found that selectively removing certain species from an ecological community changed the entire face of the ecosystem. Paine recognized the importance of these species, so he gave them a special name: keystone species.

In the decades since Paine published his idea, the **keystone species concept** has become enormously influential. In conservation efforts in particular, many projects have focused on protecting keystone species on the assumption that their loss might lead to catastrophic changes to local ecology.

However, more recently some ecologists have begun to suggest the concept can be applied too simplistically. A species that lives in many environments might be a "keystone" in one place but not in another, for instance. Despite these nuances, the keystone species concept remains an important one.

Remove the ochre sea star from rock pools and the ecosystem collapses.

2/ Shortcut: Paine performed most of his ecological studies in the intertidal rock pools on the west coast of the US. He discovered that one species of starfish – the ochre sea star – had a big impact on the ecosystem. Where the starfish is present, the rock pools are home to a variety of shellfish and other species. Paine removed the starfish from some pools and discovered that in little more than a year the diverse ecosystem collapsed: one species of mussel took over. Circumstantial evidence suggested far less dramatic effects if other predators disappeared. By eating mussels (among other things) the ochre sea star helps to maintain a diverse ecosystem. It is a keystone species.

3/ Hack: The keystone species concept suggests, bluntly, that some species are ecologically more important than others.

The extinction of any species is bad news – but the loss of a keystone species can be a disaster.

The biogenesis hypothesis

Where do living things come from?

Louis Pasteur // 1822–1895

1/Helicopter view: For about 2,000 years, European scientists thought they had a solid understanding of how new organisms appear. They simply arose spontaneously from non-living matter. Somewhat astonishingly, the idea was not finally debunked until 150 years ago – several years after **Charles Darwin** published *On the Origin of Species.*

The **theory of spontaneous generation** held that simple life arose from dust, and that even more complex organisms arose from simpler ones. For instance, **Jan Baptist van Helmont** was convinced, in the 17th century, that mice could be "grown" from a reaction between dirty shirts and grains of wheat.

Francesco Redi is widely credited with performing one of the first experiments that challenged spontaneous generation: he found that meat didn't spontaneously crawl with maggots if it was kept in a container that was inaccessible to flies.

However, despite such convincing demonstrations, a few die-hards clung to the theory of spontaneous generation even into the 19th century. In the 1860s, the influential scientist **Louis Pasteur** published the results of experiments that disproved spontaneous generation beyond all reasonable doubt. Pasteur's worked helped establish the **biogenesis hypothesis**. Simply put, this idea – now widely accepted – suggests that life comes from life (but, see: The **panspermia hypothesis**, page 204). The biogenesis hypothesis marked an important turning point in the history of science. And because of its implications, it was also a pivotal idea in the history of medicine.

Pasteur's experiments convinced many that organisms such as maggots do not appear spontaneously.

2/Shortcut: Pasteur was not the first to demonstrate that life doesn't arise spontaneously, but his formidable scientific reputation – and the convincing nature of his experiments – make his research important. He found that food would not rot if it was heat-treated in a vessel that was sealed from the surrounding air. Allow air to reach the food, though, and it began to decay in a matter of days. This outcome shows that the microbes responsible for decay must be carried in the air. They do not simply pop into existence spontaneously because of the chemical breakdown of meat.

See also //

48 The germ theory of disease, p.100

3/Hack: The biogenesis hypothesis suggests that life begets life.

In other words, all the living things now on Earth were themselves produced by other living things.

No.48
The germ theory of disease
Infectious diseases explained

1/Helicopter view: In the 1850s, cholera outbreaks were relatively common in the centre of London. **John Snow** recognized that the source of one particular outbreak was a public water pump.

On the strength of his research, the local authorities removed the pump's handle. The outbreak ended. The episode is evidence of a mid-19th century scientific revolution that led to a modern understanding of common diseases.

For centuries, European medics assumed (as the Ancient Greeks had stated) that many common diseases were triggered by noxious, inorganic air – miasma. As early as the 1540s, **Girolamo Fracastoro** had begun to argue an alternative: that diseases might come in the form of "seeds" that could be carried through the air, or spread from person to person by contact. However, Fracastoro had no way of identifying what form these hypothetical, essentially invisible seeds might take.

Although others continued to develop Fracastoro's idea, it remained a deeply unpopular concept for centuries – even after bacteria were actually discovered in the 1670s. Things began to change later in the 19th century.

Louis Pasteur was engaged in a scientific debate about putrefaction. Many believed in spontaneous generation – the idea that the microbes responsible for putrefaction simply popped into existence out of inanimate dust. In the 1860s Pasteur showed that this idea was wrong. Pasteur became convinced that the microbes responsible for putrefaction must be carried in the air – an idea that hinted at a connection between microbes and Fracastoro's disease-causing seeds.

Even so, strong evidence of a direct link between bacteria ("germs") and infectious disease had to wait for another few years. **Robert Koch** was the first person to actually demonstrate such a link through his studies of anthrax, published in 1876. The **germ theory of disease** was on the road to acceptance.

Robert Koch //
1843–1910

Koch's work provided direct evidence of a link between bacteria and disease.

2/ Shortcut: There was plenty of circumstantial evidence in favour of the germ theory of disease, but Koch provided the first direct evidence. He lived in an area where anthrax was rife. Through careful microscopic investigation, Koch found that infected animals had rod-shaped structures in their blood that were absent from healthy animals. He suggested that the structures were bacteria, and he found that healthy animals injected with bacteria-tainted blood quickly developed anthrax, hinting at a direct link between bacteria and disease.

See also //

47 The biogenesis hypothesis, p.98

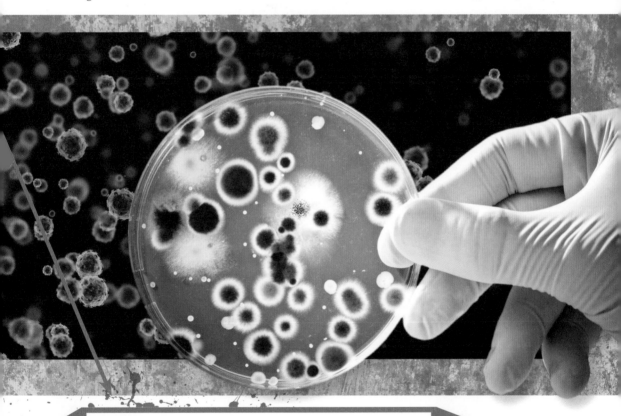

3/ Hack: Sceptical scientists took centuries to accept the germ theory of disease.

But today we take for granted the idea that infectious diseases are often caused by microbes.

The placebo effect
A medical mystery

John Haygarth // 1740-1827

1/Helicopter view: At the beginning of the 19th century many people were convinced that metallic rods – known as Perkins tractors – could provide relief from all kinds of disease. The patient simply had to rest the rods on their body and they would soon feel better. **John Haygarth** wasn't convinced: in 1801 he fashioned some fake Perkins tractors from wood, and used them to "treat" five patients. Four claimed to feel better after the treatment.

Haygarth's experiment might have been the first in medical history to formally study the placebo effect. He argued that his results showed how important a patient's hope can be in curing disease. Nevertheless, the prevailing 19th century dogma was that placebos are fraudulent. A patient's condition may improve after they are given a placebo, but this is simply because they were on the road to recovery even before they took the "treatment".

In fact, not until the 1930s did scientists begin to suspect more was going on. While trialling a vaccine for the common cold, **Harold Diehl** and his colleagues found that people given a placebo treatment responded better than people given no treatment. By the 1950s, the **placebo effect** was an accepted phenomenon.

Because the placebo effect occurs in so many different ways – there are even surgical placebos – many medical researchers assume there must be several different placebo effects, all operating through distinct pathways. It's possible, for instance, that in some cases the placebo effect is a learned response: experience teaches people to expect a therapy will work, and so their body responds accordingly. Others, including **Nicholas Humphrey**, think the placebo effect is a mystery that only begins to make sense in an evolutionary context. A complete answer to the placebo puzzle continues to elude science.

Despite two hundred years of study, scientists still don't fully understand the placebo effect.

2/ Shortcut: Humphrey argued in 2002 that the placebo effect should be understood in terms of evolution. The systems within the body that are responsible for our health have evolved to fight a disease or heal a damaged body part at the most opportune time. For instance, an animal that sprains an ankle while being chased by a lion will ignore the pain and continue running. A placebo can change the calculation: the body "thinks" that the fake treatment is real, and that it is weakening the disease. As such, the body's health systems might kick into action to take advantage of the perceived weakness – and actually fight off the disease all by itself.

See also //
1 The theory of evolution by natural selection, p.6

METALLIC TRACTORS.

3/ Hack: The placebo effect is world famous, widespread and well studied... but exactly why it exists is still a scientific puzzle.

No.50
The concept of antibiotic resistance
A predicted problem

1/Helicopter view: On 28 September 1928, **Alexander Fleming** woke up and – as he later wrote – revolutionized medicine. As the result of a mistake, one of the bacterial cultures he was growing in his lab had been contaminated with a fungus. The bacteria immediately around the fungus had died. Fleming realized that the fungus (named *Penicillium*) must produce a chemical with the power to kill bacteria. Several years later, scientists worked out how to isolate and mass-produce the chemical. Penicillin launched the antibiotic era.

But even at the dawn of the era in the early 1940s, Fleming was concerned that antibiotic-resistant bacteria might evolve. He was right to worry. Today, **antibiotic resistance** is widely considered one of the most urgent crises facing medicine.

Fleming's warning was chiefly to do with the misuse of antibiotics. For instance, if bacteria were regularly exposed to non-fatal doses of common antibiotics they might evolve new versions of genes that allowed them to withstand the drugs.

Unfortunately, misuse became common in the second half of the 20th century. Among many problems, people often took antibiotics for viral infections, even though viruses are biologically very different from bacteria and so don't respond to antibiotic treatment.

The problem was exacerbated because of another factor. At roughly the same time that Fleming was warning about antibiotic resistance, scientists were beginning to realize that bacteria have a peculiar ability to swap genes with other microbes they meet (see: **Horizontal gene transfer**, page 28). Today's researchers have discovered that this process allows drug-resistant genes to spread rapidly through microbial populations.

Fleming's work on *Penicillium* kickstarted the antibiotics revolution.

2/Shortcut: Bacteria face numerous pressures in their environment, and they are constantly evolving mechanisms to side step the problems and improve their survival odds. Antibiotics are simply another pressure. If the drugs are ever-present in low quantities, it is virtually inevitable that some bacteria in the population will evolve resistance and that this resistance will spread. No matter how effective the antibiotic, if it is misused then bacteria have an opportunity to find ways to neutralize its effects.

See also //

1 The theory of evolution by natural selection, p.6

Alexander Fleming // 1881–1955

3/Hack: Drugs misuse has helped make antibiotic resistance one of the biggest medical challenges facing the world today.

But the real tragedy is that scientists predicted the problem in the earliest days of the antibiotic era.

No.51
The neuron doctrine
The dawn of modern neuroscience

1/ Helicopter view: In 1906, **Santiago Ramón y Cajal** and **Camillo Golgi** shared the Nobel prize for medicine for their work on the structure of the nervous system – even though the two researchers vehemently disagreed on exactly what that structure was.

For centuries, scientists assumed that the nervous system operated through the movements of fluids, much like the blood system does. These "hydraulic brain" ideas fell out of fashion, particularly after **Luigi Galvani** showed in the 18th century that nerve and muscles seemed to be controlled by electrical – not liquid – signals. But exactly how those electrical signals traveled through the body was still a mystery.

Improvements in microscopy meant that, by the 1830s, it was becoming clear to scientists that living tissue was almost always constructed from cells. **Matthias Schleiden** and **Theodor Schwann** were instrumental in developing this **cell theory** – but it still seemed that the nervous system might be a strange non-cellular exception to the rule.

Golgi certainly thought it was. In the 1870s he developed a new method to "stain" nervous tissue in a biological sample so that it could be seen clearly through a microscope. His work showed the nervous

Santiago Ramón y Cajal //
1852–1934

system formed an intricate network of branch-like threads spreading out through biological tissue – but Golgi was convinced those threads fused into a single network or "reticulum". He thought the entire nervous system, from the brain downward, must form a single continuous network. The idea was known as **reticular theory**.

Ramón y Cajal began his studies of the nervous system in the 1880s. He argued that his observations showed definitively that it is, after all, composed of individual cells (later named neurons) that are not fused together. His work was instrumental in the development of the **neuron doctrine**, which ultimately became widely accepted. Neuroscience had entered the modern age.

The nervous system is composed of distinct cells that communicate at their junctions – called synapses.

2/ Shortcut: It took scientists a long time to identify individual nerve cells – neurons – partly because the cells have a very complicated structure and partly because it was difficult to find an effective way to "stain" neurons so that they were visible in their entirety under a microscope. Ramón y Cajal improved Golgi's staining technique such that the fine detail of nervous tissue was visible for the first time – detail which helped confirm that neurons are discrete elements.

See also //

52 The amyloid cascade hypothesis, p.108

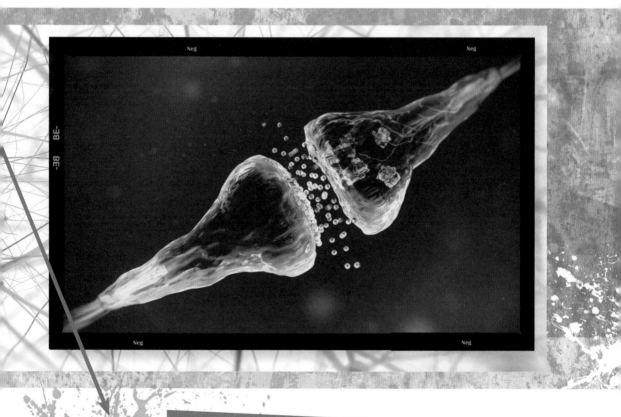

3/ Hack: The neuron doctrine states that the nervous system is built from individual, structurally complicated neurons.

It was instrumental in building a modern understanding of how the nervous system functions.

No.52
The amyloid cascade hypothesis Making sense of Alzheimer's disease

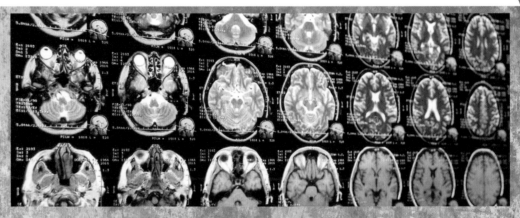

1/Helicopter view: In 1901 **Alois Alzheimer** began studying Auguste Deter, a patient at the Frankfurt Asylum who had problems with her short-term memory. Deter died a few years later at the age of 55, and Alzheimer examined her brain. He found strange protein tangles had developed around her brain cells. In 1910, Alzheimer's colleague **Emil Kraepelin** argued that Deter had died from a never-before-seen condition. In honour of his friend, Kraepelin named it Alzheimer's disease.

Neuroscientists have been debating the nature and causes of Alzheimer's disease ever since. Generally thought to be the most common form of dementia, it is associated with several features. Two distinct forms of protein build-ups – amyloid plaques and neurofibrillary tangles – often surround the brain cells of those who have died from the condition. Are they both a direct cause of the disease? Or is one (or both) simply an effect that appears as the disease develops?

In the early 1990s, **John Hardy** and **Gerald Higgins** argued that the amyloid plaques trigger the disease. They pointed to lab experiments that showed the plaques are either toxic to brain cells or increase the cells' sensitivity to the effects of other toxins. Hardy and Higgins suggested that other features of the disease appear only *after* the amyloid plaques begin to form.

Although the idea had (and still has) its critics, the **amyloid cascade hypothesis** has become the leading explanation for Alzheimer's disease. It has led to the development of therapies to reduce the formation of amyloid plaques. Some of the therapies even seem to lead to memory improvements in those with the disease.

2/ Shortcut: One key factor in the development of the amyloid cascade hypothesis was the observation that Alzheimer's is particularly common in people with Down syndrome. Those with this condition may have three – rather than the standard two – copies of chromosome 21, meaning they have an extra copy of every gene on that chromosome. This includes an extra copy of the gene that is ultimately responsible for producing amyloid plaques. As such, the cells of people with Down syndrome might make unusually large amounts of amyloid plaque, which would explain why they are particularly vulnerably to developing Alzheimer's disease.

Alzheimer's disease is associated with a confusing array of symptoms.

See also //

17 Chromosome theory of inheritance, p.38

Alois Alzheimer //
1864–1915

3/ Hack: The amyloid cascade hypothesis was an attempt to hack away at the confusing suite of symptoms associated with Alzheimer's disease.

The aim was to identify a single causal factor that could become a target for therapies.

No.53
The Hayflick limit
Why mortality begins at the cellular level

1/ Helicopter view: Humans are mortal – but early in the 20th century many scientists were convinced that our individual cells are not. In 1912, **Alexis Carrel** took heart cells from an embryonic chick and began growing them in his lab. He claimed that the cell cultures didn't die. Not until 1946 – after Carrel's death – did the "immortal" cell experiment end. The cells were simply discarded.

A few years later, **Leonard Hayflick** began to suspect there might be a problem with Carrel's immortality claim. Hayflick cultured cells as part of his work, and generally speaking the cells soon stopped growing and multiplying. In 1961, Hayflick and his colleague, **Paul Moorhead**, demonstrated that healthy human cells will divide no more than about 40 to 60 times – a figure that became known as the **Hayflick limit** of cell division.

Alexis Carrel // 1873–1944

In the 1970s it began to be clear why the Hayflick limit exists. When cells divide they replicate the DNA in their genome. But the molecule that builds the DNA strand cannot quite complete the job. It acts a bit like the slider on a zip: it binds together the two strands of the DNA molecule in its wake, but the final part of the DNA strand is beneath the molecule itself so it doesn't bind together. **Alexey Olovnikov** suggested that this means a DNA strand becomes slightly shorter every time it is replicated.

By the 1980s, **Elizabeth Blackburn** and **Carol Greider** had discovered a special enzyme – telomerase – that helps rebuild the shortened ends of the DNA strands. A 1998 study by **Andrea Bodnar** and her colleagues revealed that human cells can live beyond the Hayflick limit after all, if they are engineered to produce telomerase. Even so, few scientists view telomerase as a miracle "fountain of youth".

Most cells divide and replicate – but not indefinitely.

2/ Shortcut: Hayflick and Moorhead were sure that cell cultures do not reproduce forever, but it was possible that the cultures they studied kept dying because of a problem with their lab techniques. To rule this out, they took two types of cell – an "old" population that had divided many times and a "young" population that had divided few times – and combined some cells from each to create a mixed population. They kept an eye on the two original populations, and when the "old" one died, Hayflick and Moorhead looked at the mixed population. The "old" cells had died here too, but the "young" cells were still growing healthily. There was no problem with their lab techniques – cell cultures do die when they grow too old.

See also //

19 The double helix model, p.42

54 The cancer stem cell hypothesis, p.112

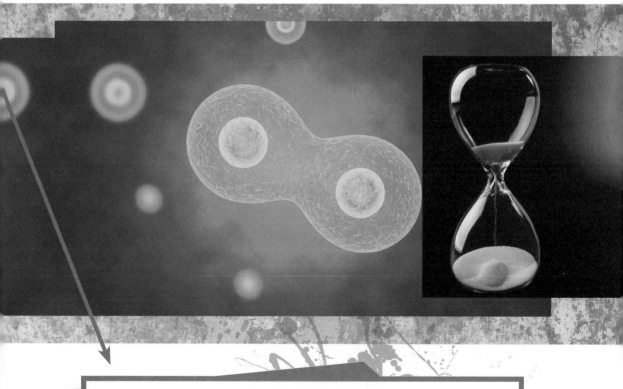

3/ Hack: The Hayflick limit puts a fundamental cap on the ability of normal human cell populations to divide.

But a few cells, including some cancer cells, seem to have found ways to avoid this limit.

No.54
The cancer stem cell hypothesis Are some cancer cells more dangerous?

1/Helicopter view: In the 1950s, **Chester Southam** and **Alexander Brunschwig** performed some experiments that, by modern standards, would be considered ethically dubious.

They took cancer cells from people with the disease and transplanted them elsewhere in the patients' bodies (the thigh or arm) to see how easily the tissue would form a new cancerous growth in an otherwise healthy region.

It proved surprisingly difficult to grow a new tumour: only if a million or more cancer cells were injected was there any chance of tumour growth. The unanswered question was: why?

By the 1990s there were two theories to explain the results. Perhaps any cancer cell can grow a new tumour but the chances of success are very low – the **stochastic theory**. Or perhaps there is a subset of very rare cancer cells that, alone, are capable of growing into a new tumour – the **hierarchy theory**.

In 1994, **John Dick** published what he thought was solid evidence favouring the hierarchy theory. Experiments with mice suggested that leukaemia could be triggered in the animals only by a subset of rare cancer cells: what came to be called cancer stem cells.

Dick's work was dismissed by many scientists, who pointed out that finding something that behaved like a cancer stem cell in just one form of leukaemia was not enough to prove that the hierarchy theory was correct. But a few years later, Dick and his colleagues reported more evidence of cancer stem cells in different types of leukaemia. By 2004, evidence of cancer stem cells in more typical "solid" tumours came to light: **Michael Clarke** and his colleagues found them in breast cancer tumours. There were still sceptics, but many scientists began to accept the hierarchy theory of cancer – also known as the **cancer stem cell hypothesis**.

Tumours might begin when a small number of cancer stem cells arrive in healthy tissue.

112

2/Shortcut: Dick was aware that most leukaemia cells seem to reproduce a finite number of times. To maintain tumour growth, he reasoned, there must be a population of special cells that can renew and reproduce almost indefinitely. In healthy tissue, cells like this are known as stem cells: perhaps, Dick thought, cancer tumours have stem cells too. When Dick and his colleagues injected human leukaemia cells into mice, a tiny proportion of the cells set up home in the rodents' bone marrow and multiplied enormously, just as cancer stem cells should. The finding convinced Dick that his idea was correct.

See also //
53 The Hayflick limit, p.110

3/Hack: Some scientists suspect cancer tumours might be sustained by a tiny subset of cancer stem cells.

If so, cancer therapies might be most effective if they can specifically target and destroy these cells.

No.55
The shape theory of olfaction   How do we smell?

1/Helicopter view: In 1870, **William Ogle** speculated on the mechanism responsible for our sense of smell. At the time it seemed clear that the senses of sight and hearing involved light and sound waves, respectively. Perhaps, he suggested, the sense of smell also involved waves or vibrations of some kind.

Malcolm Dyson advanced the idea in the 1920s, and many scientists began to think that the smell of a molecule must be tied in some way to the distinctive way the bonds between its constituent atoms vibrated. But gradually, the idea fell out of favour, largely because experiments aimed at demonstrating the link between smell and molecular vibrations failed.

At the same time another model emerged to replace this **vibration theory of olfaction**. In the mid-1940s **Linus Pauling** emphasized the important role that a molecule's physical *shape* played in the way it reacted: at the end of the decade **Robert Wighton Moncrieff** suggested that this property of shape might be at the root of the sense of smell.

Richard Axel // b. 1946

The idea assumed smell was based on a "lock and key" system: inside the nose there must be a system of receptors onto which odour molecules in the air attach in order to be smelled.

Moncrieff's **shape theory of olfaction** received a significant boost in the 1990s, when **Linda Buck** and **Richard Axel** identified what appeared to be smell receptors inside the rat nose, and found they behave in the way that the shape theory assumes they should. Subsequently, the shape theory became the dominant model. However, an updated vibration theory of olfaction emerged in the 1990s too – its advocates say it is too early to close the book on the science of our sense of smell.

Only recently has our sense of smell begun to make complete sense.

2/Shortcut: Despite all of the progress made in understanding the nervous system, the neural receptors that respond to smells were not clearly identified until 1991. Buck and Axel looked at the way genes operate inside the thin layer of skin inside the rat nose and found that a "family" containing hundreds of related genes was particularly active. Each of these genes produces a distinct protein molecule. Buck and Axel thought it was likely that these molecules are smell receptors – and the fact there are so many of them suggests the sense of smell is complicated, with odour molecules in the air each binding to a specific receptor. This was compatible with Moncrieff's shape theory of olfaction.

See also //

51 The neuron doctrine, p.106

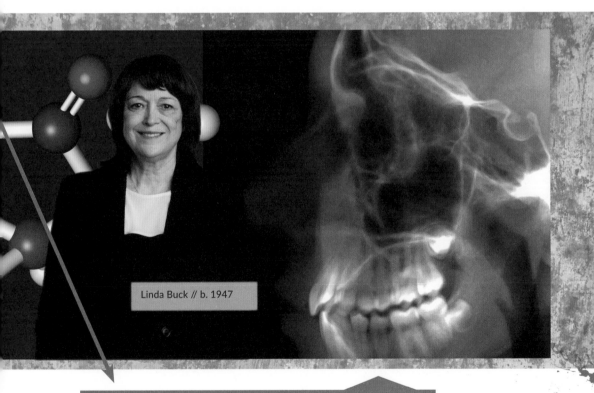

Linda Buck // b. 1947

3/Hack: Of the traditional five senses, the sense of smell is the most difficult to explain.

But most scientists think odour molecules are "keys" that fit specific smell receptor "locks" in our nose.

No.56
Newton's law of universal gravitation
The most famous apple in science

1/Helicopter view: Many of the top scientific minds of the 17th century struggled to understand the nature of gravity – then **Isaac Newton** had an encounter with an apple and made a key breakthrough.

The story of Newton's apple has attained urban myth status, but there's reason to believe it is based in fact – it is mentioned in a document written by Newton's contemporary, **William Stukeley**.

No one can be completely sure what went through Newton's mind when he saw the apple fall, but the assumption is that he realized the fruit had moved from a state of rest (attached to the branch) to a state of motion (plummeting ever faster toward the ground). This implied that a mysterious force must be acting to accelerate the apple on its journey earthward – this force is what we recognize as Earth's gravity.

Then came what is widely recognized as Newton's flash of brilliance. If Earth's gravity acts on an apple growing just above the planet's surface, might it not also act on objects much further away? Newton suggested it might. By Newton's day it was clear that the Moon is in motion around the Earth. Newton calculated that it is moving with just enough speed to balance the force of Earth's gravity: it orbits the Earth because it isn't moving fast enough to escape Earth's gravitational pull and fly off, or slow enough to succumb entirely to Earth's gravity and fall into our planet.

And why stop there? Newton argued that his idea could also explain why Earth is in motion around the Sun, for instance. In fact, he suggested that gravity was *universal*: every object in the Universe must be under the gravitational influence of every other object – although the further the distance between two objects, the smaller the influence. The **law of universal gravitation** became enormously influential.

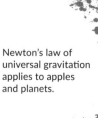

Newton's law of universal gravitation applies to apples and planets.

2/ Shortcut: Imagine Newton picked up his famous apple and threw it. It might travel a short distance before it hits the ground. An athlete with a better throwing arm could fling the fruit much further – ten times as far, say. Now imagine an unfeasibly strong athlete: this person can throw the apple so far it travels all the way around the planet. In fact, if the athlete gets his throw just right, the apple will still be at the same height it was when it left his hand even after traveling all the way around the Earth. It will carry on flying around the planet (or it will if we simplify things by assuming that the apple will not be slowed down by the air's friction). The apple is in orbit. Through this sort of reasoning, physicists accepted that Newton's law explains why the planets move around the Sun.

See also //

57 Newton's laws of motion, p.118

68 The general theory of relativity, p.140

$$F = G\ \frac{m_1 m_2}{r^2}$$

3/ Hack: Every object – whether a planet or a person – is pulling every other object in the Universe towards it.

The strength of that gravitational pull depends on the object's mass and its distance from the second object.

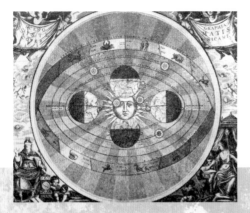

Newton's laws of motion
How the heavens lost their mystery

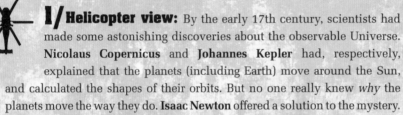

1/Helicopter view: By the early 17th century, scientists had made some astonishing discoveries about the observable Universe. **Nicolaus Copernicus** and **Johannes Kepler** had, respectively, explained that the planets (including Earth) move around the Sun, and calculated the shapes of their orbits. But no one really knew *why* the planets move the way they do. **Isaac Newton** offered a solution to the mystery.

Newton's three laws of motion explain how an object changes its speed and direction of movement when acted on by a force. Probably the most famous of Newton's three laws is the third: *for every action there is an equal and opposite reaction.* Newton illustrated the idea with a horse pulling a stone tied to a rope. When the rope is taut, the stone is pulled toward the horse – but the horse is also pulled back toward the stone to exactly the same extent. In other words, the stone has an impact on the horse's forward movement – which explains why the animal might struggle to shift the stone.

Newton's laws of motion help explain the movement of objects on Earth. More significantly, Newton realized they could also explain why the planets move the way they do.

By the time Newton outlined his laws of motion in the 1680s, he had already demonstrated that the force of gravity acts on everything in the Universe. Newton combined his understanding of gravity with his emerging understanding of the way that his laws of motion influence objects. Suddenly the paths those planets take through the sky made sense.

Scientists quickly accepted Newton's ideas, and they remain a core element of physics to this day. After Newton, the Universe never seemed quite so mysterious and unknowable.

Copernicus helped change our view of the Solar System – but Newton explained why planets move as they do.

2/ Shortcut: A car – and its driver – traveling at a constant speed down a road will remain at that speed unless another force intervenes. If the car slams into a wall, the wall provides that intervening force. It pushes back against the car and the vehicle loses speed abruptly. Unfortunately, the driver – who forgot to fasten his safety belt – does not. Because he hasn't actually hit the wall, he continues to move at speed, tumbling out of the driver's seat and smashing through the windscreen. He is a victim of Newton's laws of motion.

See also //
56 Newton's law of universal gravitation, p.116

Nicolaus Copernicus//
1473–1543

3/ Hack: Newton's laws of motion are central to the scientific understanding of the observable Universe.

They argue that all objects, from people to planets, move and change speed and direction in essentially the same way.

No.58

The corpuscular theory of light
What exactly *is* light?

Pierre Gassendi // 1592–1655

1/ Helicopter view: Intellectual thinkers have been debating the nature of light for at least 2,500 years. A modern understanding began to emerge in the 17th century. **Pierre Gassendi** played an important role in the process.

By the 1640s, Gassendi was convinced that the world and everything in it was constructed from very tiny indivisible particles, or "corpuscles". In Gassendi's opinion, it was not just *physical* objects that were built from corpuscles – sound, heat and *light* were carried from one location to another in corpuscular form too.

Isaac Newton read Gassendi's work and liked many of its ideas. By the 1670s he was convinced that light was best described as a stream of particles – rejecting the **wave theory of light**, an alternative model favoured by researchers including **René Descartes** and **Robert Hooke**.

Newton published a formal version of his ideas in 1704. His book, *Opticks*, also included details of his experiments into the nature of colour, and quickly became a key text. For the next century, most scientists accepted the **corpuscular theory of light** – but ultimately physicists realized light was far stranger (see: **The complementarity principle**, page 158).

Newton's experiments with light convinced him it is carried by tiny particles.

2/Shortcut: Newton had good reason for envisioning light as a stream of tiny particles: the model helped explain some of the observations he had made while experimenting with light. Reflection, in particular, was a key consideration. A ray of light reflecting off a mirror moves along the same sort of trajectory that would be expected of a ball bouncing off the mirror. Newton also thought it was significant that light does not appear to travel around corners in the same way that sound can (which he assumed *was* carried by waves). Observations of this sort made sense to Newton only if light was carried by particles, not waves.

See also //
59 The wave theory of light, p.122

3/Hack: Newton argued that light was carried in the form of impossibly small particles.

Partly because of Newton's formidable reputation, this corpuscular theory of light became very popular.

No.59
The wave theory of light
An idea that kept rising and falling

1/Helicopter view: In 1803, **Thomas Young** reported the results of a deceptively simple experiment – and changed the way scientists thought about light. His findings were so startling that they overturned a century-long consensus that had begun with the great **Isaac Newton**.

Put simply, Young's "double-slit experiment" demonstrated that light behaves in a way that can only be explained by assuming it ripples through the air as a series of waves. Prior to his experiment the dominant view – popularized by Newton – was that light rays actually contain a stream of tiny ball-like particles (see: **The corpuscular theory of light**, page 120).

In truth, Young's experiment was merely reviving views that had been expressed by scientists working more than a century earlier. **René Descartes**, for instance, had argued that light behaved like a wave in the 1630s.

Later in the 17th century, **Christiaan Huygens** expanded on Descartes's work. He described rays of light as waves and found that his model could explain not only how light rays *reflect* off surfaces but also how they *refract* – or "bend" – as they travel from one medium (for instance, air) into another (such as water).

Thomas Young // 1773-1829

However, although Huygens's model of light was arguably at least as persuasive as Newton's, it was the latter who most strongly influenced the way 18th century scientists thought of light. This may have been more about Newton's general reputation than because the corpuscular theory was superior to the wave theory.

In any case, Young's experiment at the turn of the 19th century helped trigger a scientific change of mind. A few years later, **Augustin-Jean Fresnel** built on Huygens's ideas and showed how they could explain Young's results. The **wave theory of light** became the dominant view – or at least it did until the beginning of the 20th century (see: **The photoelectric effect**, page 152).

Young's experiment seemed to make it clear that light moves through air in a series of waves.

2/Shortcut: Young shone an even light onto an opaque screen in which two tiny slits had been made. According to the corpuscular theory of light (and simple common sense) two tiny slivers of light should then have been visible on the wall behind the screen. In fact, Young saw several light bands separated by dark bands in a barcode-like arrangement. He realized that the light had been diffracted – spread out – as it traveled through the narrow slits in exactly the same way water waves are when they squeeze through a narrow opening. The light waves spreading out from the two slits interacted and interfered with each other (again, as water waves would) to make light and dark bands on the wall.

See also //

80 The electron double-slit experiment, p.164

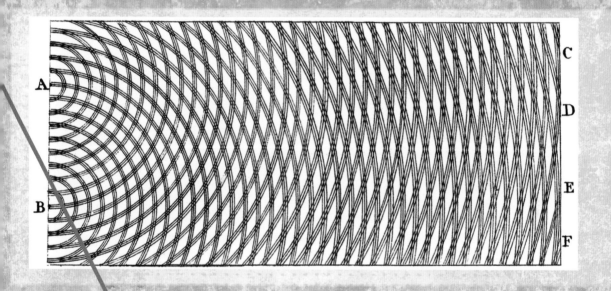

3/Hack: Newton had helped convince the scientific community that light was carried in particle form.

Young's simple experiment was so convincing that it revived the moribund theory that light moved as a wave.

No.60
Maxwell's equations
Seeing light in a new light

1/Helicopter view: A series of events during the 19th century led to a revolutionary new understanding of light – one that ultimately paved the way for our modern technological world. Surprisingly, the story begins with a magnetic compass.

During a lecture he gave in 1820, **Hans Christian Oersted** noticed that the needle of his compass deflected slightly when he passed an electric current through a nearby wire. He published his findings, and other physicists quickly began probing a possible connection between electricity and magnetism – **Michael Faraday** among them.

Faraday soon discovered he could generate an electric current in a wire simply by moving a nearby magnet. This demonstrated that Oersted's observed link between electricity and magnetism was a two-way street. Electricity and magnetism were, in effect, two sides of the same coin.

But it still wasn't clear exactly *how* magnets could influence electric currents and vice versa. Many physicists simply assumed some mysterious direct and instantaneous connection between the two. Faraday did not like this idea: he thought there had to be a mechanism by which the two "communicated". He argued for the existence of invisible "lines of force" travelling through a medium – air, for example – at a finite speed and influencing both the electric and magnetic properties of objects.

The idea seemed vague to most numerically minded 19th century physicists – but not to **James Clerk Maxwell**. In the 1860s, he published his mathematical investigations into the idea. He concluded that Faraday was essentially correct. **Maxwell's equations** formally unified electricity and magnetism in a single force – electromagnetism. They also strongly suggested light was an electromagnetic phenomenon. Maxwell's astounding insights were accepted by other scientists, and paved the way for later researchers to manipulate light for use in applications as diverse as medical imaging and global communications.

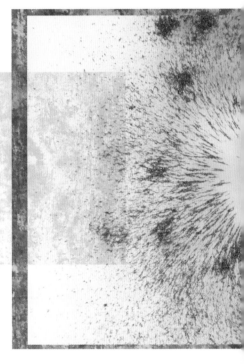

Iron filings scattered on a magnet trace out a magnetic field – similar fields can be generated with electricity.

2/Shortcut: Maxwell argued that space is permeated by an invisible medium – the **"ether"** – through which electromagnetic phenomena travel as waves. These electromagnetic waves carry information between objects, explaining (for instance) how a moving magnet could generate an electric current in a wire. Maxwell's calculations predicted that these waves travel extraordinarily quickly – in fact, at a value similar to the calculated speed of light. This implied that light *was* an electromagnetic wave (see: **The wave theory of light**, page 122). In the 1880s, **Heinrich Hertz** provided strong evidence in favour of Maxwell's concept.

See also //

65 The ether hypothesis, p.134

66 The special theory of relativity, p. 136

James Clerk Maxwell // 1831–1879

3/Hack: At the dawn of the 19th century scientists knew a lot about electricity, magnetism and light, but nobody dreamed the three were intimately linked.

Maxwell's equations made that connection – and changed the world.

Boyle's law
The beginning of the industrial revolution

Robert Boyle // 1627–1691

1/ Helicopter view:
The 17th century was a time of scientific revolution as the greatest minds of the age began to question assumptions handed down from the ancient world. Famously, **Aristotle** is said to have stated that "nature abhors a vacuum", arguably suggesting that it was physically impossible to generate vacuums. By the 1640s there were signs that Aristotle was incorrect.

It may have been **Evangelista Torricelli** who first created a vacuum in an experiment in 1643. Others followed, including **Blaise Pascal** and **Otto von Guericke**. In one experiment, von Guericke placed two copper hemispheres together to create a hollow globe, and sucked out the air trapped inside. Even two teams of horses could not pull the two pieces of copper apart.

This experiment suggested that standard atmospheric pressure must exert a huge amount of force on objects – enough to keep the two pieces of copper locked together despite every effort to separate them. **Robert Boyle** heard about the experiments and began his own investigations with his assistant, **Robert Hooke**. By the mid-1660s, Boyle had discovered a relationship between the pressure imposed on a quantity of air and the volume that the air occupied. This became known as **Boyle's law**.

Denis Papin began working with Boyle in the 1670s, exploring other implications of Boyle's law. By now it was clear that air pressure could also be increased by raising the temperature of air trapped inside a sealed container. Papin used the idea to invent an early forerunner of the pressure cooker. He later added a safety valve to prevent a potentially dangerous explosion. As Papin watched the valve moving up and down to release pressure, he realized heat and pressure could be used to move objects. By the end of the 17th century the first steam engines had been built, and the industrial revolution moved a step closer.

Arguably, the invention of the steam engine was a direct result of Boyle's work on gases.

2/Shortcut: Boyle assumed that gas has elastic properties: like a spring, gases could be squashed, but would bounce back to their original volume once the pressure on them had been released. It was while exploring this spring-like property that he noticed a pattern. When he squeezed a bubble of air trapped in a sealed tube, he found that the pressure he applied – multiplied by the volume that the squeezed air bubble came to occupy under that pressure – always gave roughly the same number. In other words, as pressure went up, volume went down – and vice versa.

See also //

63 The second law of thermodynamics, p.130

70 The kinetic theory of gases, p.144

3/Hack: Boyle's law states that pressure and volume are inversely proportional if temperature remains constant. The law was instrumental in building the first steam engines.

In effect, Boyle's law helped kick-start the industrial revolution.

No.62
The first law of thermodynamics
The ultimate form of conservation

1/ Helicopter view: James Joule had practical reasons for exploring the world of physics. As a 19th century brewer in the north of England, he was keen to work out how the new technologies of the industrial revolution could be best used to maximize his profits. His studies into heat would help revolutionize science too.

For decades before Joule, most physicists had argued that heat was a fluid of sorts that naturally flowed from hot regions to colder ones, in much the same way that water flows from a higher point to a lower point. A central assumption of this **caloric theory** was that heat could never be created or destroyed.

Joule's experiments in the 1840s strongly questioned this assumption. He argued that a simple turbine would, in fact, generate "new" heat as it mechanically turned. Joule's peers in the United Kingdom were sceptical of this claim. But scientists elsewhere in Europe were arriving at a similar conclusion. A few decades earlier, **Sir Benjamin Thompson, Count von Rumford,** had noticed that when a machine drilled the bore in a brass cannon, it generated heat. **Julius Robert von Mayer** and **Hermann von Helmholtz** made similar observations.

James Joule // 1818–1889

Helmholtz realized that such experiments hinted at a universal principle. Heat certainly can be created and destroyed, but in a broader sense, he argued, energy cannot be: it simply changes the form it takes (but see: $E = mc^2$, page 138). Helmholtz's conclusion was highly influential. Within a few years the **law of conservation of energy** was widely accepted. It is now recognized as the **first law of thermodynamics**.

As a turbine spins, it generates heat through friction.

2/ Shortcut: Joule's most famous experiment into the nature of heat was a relatively simple one. He took an insulated barrel filled with water and containing a turbine, which was attached via a string to a weight. When Joule dropped the weight, the turbine in the barrel turned. As it spun, Joule registered a small but significant increase in the temperature of the water in the barrel – what scientists now recognize as heat from friction of the turbine blades cutting through the water. Joule had "created" heat. But crucially, he had not created energy. In suspending the weight from a string, Joule had filled it with "potential" energy. As it dropped this energy dissipated – some of it heated the water.

See also //

63 The second law of thermodynamics, p.130

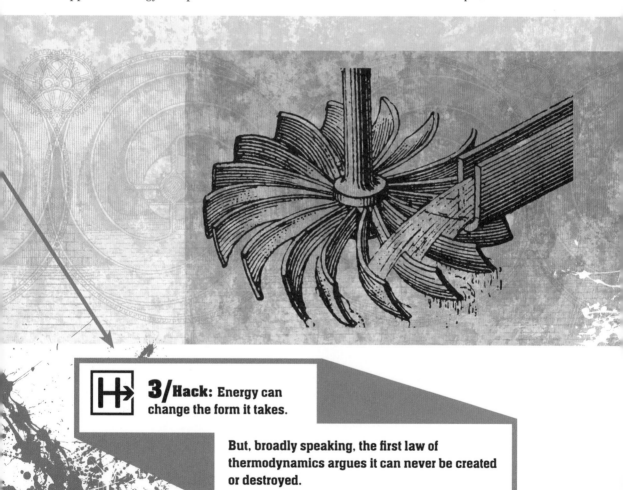

3/ Hack: Energy can change the form it takes.

But, broadly speaking, the first law of thermodynamics argues it can never be created or destroyed.

No.63

The second law of thermodynamics
Why quantity trumps quality

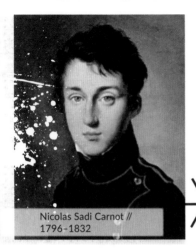

Nicolas Sadi Carnot //
1796-1832

1/Helicopter view: Scientific breakthroughs of the 17th century led to a tangible benefit: the steam engine (see: **Boyle's law**, page 126). But early steam engines were painfully inefficient. In the 1820s, **Nicolas Sadi Carnot** explored ways to remedy the problem.

Some engineers had suggested using fluids other than steam in a bid to improve engine efficiency. Carnot, however, argued that the key variable was temperature – specifically, the temperature difference between the engine's hot furnace and the relative cool of the surrounding environment. Carnot said an engine takes advantage of the flow of heat from the furnace to this environment to move its pistons – to do "work".

Carnot also argued that a theoretically ideal engine does this so well that all the heat – the useful energy – in the furnace is either converted into work, or remains in a potentially useful state.

Carnot was 25 years ahead of his time. The **first law of thermodynamics** was yet to be formulated. However, physicists of the 1850s – particularly **Rudolf Clausius** and **William Thomson** (later **Lord Kelvin**) – realized that his arguments offered the earliest insight into another fundamental thermodynamic principle. Put simply, they argued, there is a one-way direction to processes like the running of an engine, and an inevitable decline in useful energy with time. This is because nothing in the real world can match the efficiency of Carnot's ideal engine. During the operation of a real engine – or anything (including a living organism) powered by energy – some potentially useful energy will be converted into a useless form. The total *quantity* of energy might remain the same but, over time, its *quality* drops. By the 1860s Clausius had begun labelling this inevitable increase in the proportion of useless energy with time an increase in "entropy". These ideas became the **second law of thermodynamics**.

The second law of thermodynamics puts a cap on the efficiency of engines.

2/ Shortcut: A thermodynamic system – whether a steam engine, a living organism, or the observable Universe – may exploit the movement of (potentially useful) energy from hot to cold regions. But these systems do this in a way that converts at least some of that useful energy into a useless form. Most obviously, a steam engine converts some energy into frictional heat as its pistons pump. There is an inevitable increase in useless energy (or entropy) with time.

See also //

62 The first law of thermodynamics, p.128

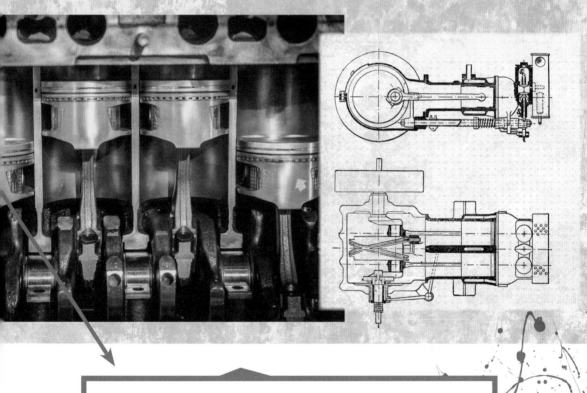

3/ Hack: In any thermodynamic system, there is a gradual loss of useful energy.

Unless you "cheat" by supplying the system with a new source of useful energy, it will eventually stop working.

No.64
The concept of absolute zero

How low can you go?

-273.15°C

1/ Helicopter view: The discoveries that 17th century researchers made about gases helped trigger the industrial revolution (see: **Boyle's law**, page 126). But there was much more still to learn about gases and about the fundamentals of nature.

This story arguably begins with **Guillaume Amontons**. Early in the 18th century, he exploited the observation that a gas changes volume with temperature to design a new type of thermometer – one in which the "springiness" of warm air forced mercury to rise up a tube. Amontons used his thermometer to make a bold prediction: air would, in theory, lose all of its energetic springiness if it were cooled to a temperature of -240°C (although he was working long before the Celsius scale was adopted).

Few of Amontons's contemporaries were interested in this work – but in the 1770s **Johann Heinrich Lambert** returned to the idea. With improved equipment he refined Amontons's prediction: theoretically, he argued, air could be cooled to -270°C before losing all of its energy.

By the early 19th century, science had moved on. Many were arguing that all gases – not just air – expanded or contracted with temperature in a broadly similar way. In the 1830s, **Émile Clapeyron** developed this concept into the **ideal gas law**. This argues that a theoretically "ideal" gas will continue to change volume with temperature in the same way at all temperatures (real gases do not do this: they eventually condense into liquids or solids as they are cooled).

In 1848, **William Thomson** (later **Lord Kelvin**) took inspiration from ideas like this to argue for an absolute temperature scale based on "ideal" thermodynamic principles rather than a relative scale based on the behaviour of specific materials. The **concept of absolute zero** was born, and soon became widely accepted. Today, it has been refined to -273.15°C.

William Thomson //
1824–1907

2/ Shortcut: Inside Amontons' gas thermometer, the level of mercury in a glass tube rose and fell depending on the temperature of a trapped air bubble. At water's boiling point (100°C) the mercury rose to a level of "73" on Amontons' scale (measured in inches). But even at water's freezing point (0°C) the mercury stood at 51.5: air was still (using Amontons' terminology) very "springy". Assuming that air will continue to behave as a gas as it is cooled (in other words, assuming air is what modern physicists would call an "ideal gas"), Amontons could easily calculate that the temperature would have to drop to -240°C before air would have lost all of its spring (and energy) and the mercury would stand at zero on his scale.

Absolute zero – the theoretically lowest temperature in the observable Universe.

See also //

70 The kinetic theory of gases, p.144

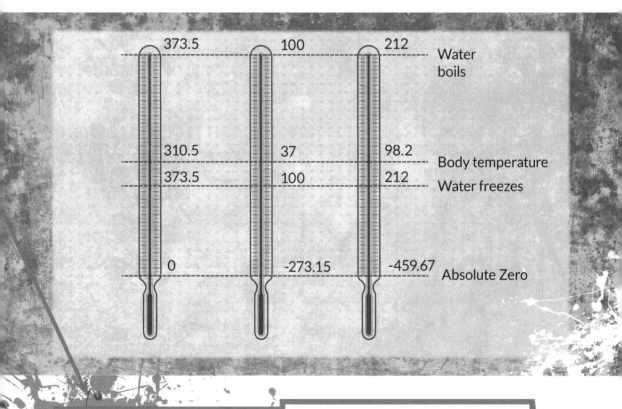

373.5	100	212	Water boils
310.5	37	98.2	Body temperature
373.5	100	212	Water freezes
0	-273.15	-459.67	Absolute Zero

3/ Hack: Absolute zero is a universal limit.

It is a temperature so low that, using thermodynamic principles, it is impossible to actually reach.

No.65
The ether hypothesis
What does light travel through?

 1/ Helicopter view: Physicists of the late 19th century were reasonably confident that they understood the nature of light. Thanks to the spectacular work of **James Clerk Maxwell**, it was widely accepted that light was a form of electromagnetic radiation, traveling as waves (see: **Maxwell's equations**, page 124). There was just one remaining question: exactly *what* were those waves traveling through?

Typically, waves move through some sort of medium. Most obviously, ocean waves move through water. The consensus in the late 19th century was that light waves move through something called the "ether" that permeates the entire observable Universe.

By the 1880s, building on the work of earlier physicists including **Augustin-Jean Fresnel**, many assumed that the Earth ploughed through this ether as it orbited the Sun, constantly changing its sense of movement relative to the universal "ether wind".

Albert Michelson reasoned that, if this was the case, the ether wind should have detectable influences on the speed of light – in much the same way that an aeroplane's speed changes if it is flying into or with the prevailing air currents in the atmosphere. He designed an experiment to detect these differences. But he found no evidence that the speed of light changed in line with the ether wind idea.

Albert Michelson // 1852–1931

This result astonished other physicists. Later in the 1880s, Michelson worked with **Edward Morley** to build a better, high-precision version of his experiment to double-check the finding. But even this equipment failed to uncover any evidence of the ether wind. The **ether hypothesis** was in real trouble.

Physicists once assumed that a mysterious "ether wind" swept across Earth as it orbits the Sun.

2/Shortcut: Michelson reasoned that a beam of light leaving a fixed point (A) and traveling in the same direction as the ether wind should move faster, just as an aeroplane flies faster with a following wind. He further reasoned that if the light was then reflected so it began travelling *into* the ether wind back to point A, it would move more slowly. Crucially, Michelson knew from mathematics that light should slow down to a *greater* extent on this return journey than it sped up on the outward journey: if the ether wind exists, the light should return to point A slightly later than it would do if there was no wind present. However, Michelson (and Morley) found that light always arrived back at point A right on time. There was no ether wind – perhaps because there was no ether.

See also //

66 The special theory of relativity, p.136

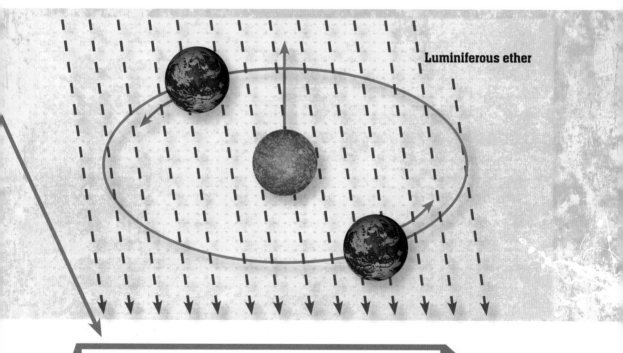

Luminiferous ether

3/Hack: Going back as far as the 17th century, some scientists had assumed light travels through an "ether" permeating the Universe.

In the 1880s, Michelson and Morley effectively destroyed this longstanding idea.

No.66
The special theory
of relativity Rethinking reality

1/Helicopter view: By the 1880s, many physicists assumed light waves traveled through an invisible "ether" that swept across the planet. They thought that the ether wind should have subtle but detectable effects on the speed of light at Earth's surface – but **Albert Michelson** and **Edward Morley** found that the speed of light on Earth's surface was invariable. A final attempt to shore up the **ether hypothesis** would lead to an entirely new way of thinking about the Universe.

Within a few years of the Michelson-Morley experiment, both **George Francis FitzGerald** and **Hendrik Lorentz** argued that light does travel slightly slower when moving into the ether wind, as assumed. But, they said, physical objects and the rate clocks tick are also slightly distorted by the wind. This, they suggested, explained the Michelson-Morley result: all of the distortions cancelled out, so light speed seemed to be constant.

Hendrik Lorentz // 1853–1928

This was not a popular idea. It seemed completely untestable. In 1905, though, **Albert Einstein** turned a modified version of the concept into a testable theory.

Einstein was impressed with **James Clerk Maxwell**'s work on electromagnetism. He realized that **Maxwell's equations** implied that the speed of light through a given medium was always the same (explaining the Michelson-Morley result). Einstein urged physicists to accept this as a core assumption. Doing so suggested FitzGerald and Lorentz were, in a sense, correct – but Einstein argued it was space and time themselves (that is, the "ether") rather than physical objects that are distorted in order to maintain a constant speed of light. Einstein's idea had a solid mathematical foundation and made testable predictions which were later confirmed.

Einstein's **special theory of relativity** did not exactly disprove the ether hypothesis, but it rendered the ether so malleable that it was impossible to confirm its existence: as such, physicists abandoned the ether concept.

Lorentz and Einstein's work helped point toward a new understanding of time, space and reality.

2/Shortcut: Maxwell's equations implied that the speed of light in a vacuum is constant. But imagine a spacecraft moving at high speed away from the Sun: logically, the spaceship crew gazing back toward the Sun will be looking at sunlight traveling at the speed of light plus the speed of their spacecraft. The special theory of relativity predicts they do not. As the spacecraft picks up speed, its onboard clocks automatically begin to slow down (relative to observers on a nearby stationary space station). When the spacecraft crew measures the speed that light is leaving the Sun, they are relying on their slow timepieces to do so, and they find that the sunlight is still traveling at the speed of light. This prediction has now been confirmed experimentally.

See also //

60 Maxwell's equations, p. 124

3/Hack: It is natural to assume that time and space are fixed and invariable properties of reality, but special relativity argues they are not.

Time can run more slowly and space can shrink – and the greater the difference in speed between two observers, the greater the effect.

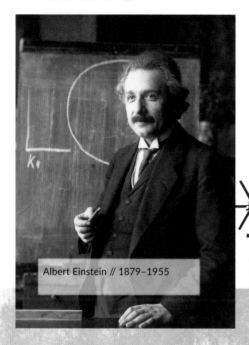

Albert Einstein // 1879–1955

E=mc²
The world's most famous equation

1/Helicopter view: In 1905, **Albert Einstein** published what would come to be known as his **special theory of relativity** – and changed the way physicists think about the Universe. Einstein later suggested that the single most important consequence of special relativity is an idea represented by an elegantly simple equation: $E = mc^2$.

$E = mc^2$ is perhaps easiest understood in its historical context. Little more than one hundred and fifty years ago, most physicists were convinced that heat can never be created or destroyed – an idea known as the **caloric theory**. Today it is clear even to non-physicists that this "conservation of heat" principle is not correct: a car engine heats up as it operates – it creates "new" heat.

However, it was only because of the careful work of **James Joule** and others in the middle of the 19th century that physicists were persuaded to abandon the conservation of heat principle (see: **The first law of thermodynamics**, page 128). By the end of that century, most scientists accepted that heat is merely one form of energy, so it *can* be created or destroyed by conversion into another form of energy. But energy itself, they thought, *does* obey a conservation principle: it cannot be created or destroyed.

Crudely speaking, what Joule's experiments of the 1840s did to the conservation of heat principle, Einstein's special theory of relativity did to the conservation of energy principle. It showed that the idea was not quite correct. Einstein argued that energy and mass are fundamentally the same thing. This was a bold prediction, but one that physicists later verified through experiments.

Energy and mass are intimately linked, as explored by physicists at CERN near Geneva.

2/Shortcut: Einstein's equations made clear that energy is actually just one of two forms of "mass-energy". He effectively showed that energy (E) can *become* mass (= m), and mass can *become* energy. In our daily existence we aren't really aware of this mass-energy equivalence because enormous amounts of energy are required to generate even tiny amounts of "new" mass. This explains the "c^2" part of the equation. Physicists use "c" to represent the speed of light, which is a very large number. It becomes much, much larger when it is multiplied by itself.

See also //
66 The special theory of relativity, p.136

SHFST - No.

3/Hack: $E = mc^2$ is a world-famous equation that carries a revolutionary idea:

energy and mass are two sides of the same coin.

No.68
The general theory of relativity
Why gravity is like a rubber sheet

1/Helicopter view: Right at the start of the 20th century, **Albert Einstein** changed the scientific understanding of time, space and energy (see: **The special theory of relativity**, page 136). But his ideas worked only in a "special" case that assumed objects were moving at constant, uniform speeds. Did relativity still apply in a *general* sense for objects changing speed? In 1915, Einstein concluded that it did. In doing so, he revolutionized the scientific understanding of gravity.

A key feature of what became Einstein's **general theory of relativity** is the "principle of equivalence". Einstein argued that the laws of nature, as applied to an object that is accelerating, are basically identical to the laws of nature at work on a stationary object in a gravitational field (Earth's gravitational field, for example).

$$G_{\mu\nu} + \Lambda g_{\mu\nu} = \frac{8\pi G}{c^4} T_{\mu\nu}$$

As he continued to explore this principle, Einstein's ideas began to clash with traditional expectations of gravity. **Isaac Newton** had considered gravity to be a mysterious force of attraction between anything (and everything) in the observable Universe. Einstein realized that in his model – which already allowed for the distortion of time and space – gravity could be understood in a different way.

Put simply, Einstein argued that gravity *is* the distortion of time and space. Galaxies and stars distort "spacetime" in much the same way that a bowling ball placed on a stretched rubber sheet distorts the material. Spacetime distortions, in turn, influence the behaviour and distribution of mass (and energy; see: **E = mc²**, page 138). Likewise, the path a marble takes as it rolls across the rubber sheet is influenced by the distortion created by the bowling ball: it is "pulled" toward the bowling ball, just like gravity pulls objects toward one another in spacetime.

The general theory of relativity made bold but testable predictions: as those predictions were confirmed, the idea's popularity grew.

Einstein's equations pointed toward a new understanding of gravity.

2/ Shortcut: An astronaut in a space capsule that is accelerating at *exactly the same rate* as Earth's gravitational pull will see objects drop to the ground in the same way they do on Earth's surface: acceleration and gravitational fields are *equivalent*. Now, imagine the capsule begins accelerating extremely quickly. The astronaut opens and shuts a window, allowing in a horizontal shaft of light from a nearby star. The shaft of light will appear to *bend* as it travels across the capsule because of the craft's extreme acceleration. Equivalently, said Einstein, rays of light will bend if they travel through an extremely strong gravitational field like the one near the Sun – his prediction was confirmed in 1919.

See also //

56 Newton's law of universal gravitation, p.116

3/ Hack: Einstein's general theory of relativity suggests matter distorts space and time, and that these distortions in turn influence the behaviour of matter and energy.

These distortions in spacetime are gravity.

John Dalton // 1766-1844

Dalton's atomic theory
Some things just do add up

1/ Helicopter view: The idea that matter might be constructed from tiny atoms stretches back millennia. But not until the turn of the 19th century did the idea begin to gain a solid, scientific standing. One of the key figures involved in this transition was **John Dalton**.

While Dalton was in his twenties, two important scientific laws were being formulated. First, in the 1780s, **Antoine Lavoisier** realized that chemical reactions neither created nor destroyed mass: the weight of the reactants at the beginning of the experiment was *identical* to the weight of the products at the end. This became the **law of conservation of mass**.

A decade later, in the 1790s, **Joseph Proust** made another important discovery. He was studying chemical compounds – substances containing elements of more than one kind. He realized that when they are broken down into their constituent elements, and those elements are weighed, the relative weights are always in proportion to one another. For example, approximately 11 per cent of the mass of a glass of water is hydrogen and 89 per cent is oxygen. Pour out some of the water and examine it again, and 11 per cent of the remaining mass is still hydrogen and 89 per cent is still oxygen. This became known as **Proust's law** or the **law of definite proportions**.

Dalton argued that these two laws could be combined to generate a deeper understanding of nature; one that recognized matter must be composed of tiny – but discrete and apparently indivisible – atoms. Dalton was working at a time when it was impossible to actually *observe* atoms using scientific equipment, so his ideas were based largely on theoretical principles. Solid though his **atomic theory** was, the debate on the existence of atoms would continue even into the early 20th century (see: **Einstein's theory of Brownian motion**, page 146).

Dalton's careful observations and arguments made a convincing case for the existence of atoms.

2/ Shortcut: Dalton made an important addition to Proust's law during the early 19th century. He realized that, when two elements can combine in more than one way to produce a compound, there was still a pattern in the proportions involved. For instance, carbon can react with oxygen to produce two distinct gases (carbon dioxide and carbon monoxide). Dalton recognized that, starting with a certain mass of carbon, exactly twice as much oxygen is needed to generate carbon dioxide as is needed to generate carbon monoxide. Patterns like this made sense to Dalton only if matter was composed of, in **Isaac Newton**'s words, "solid, massy, hard, impenetrable, movable particles".

See also //

70 The kinetic theory of gases, p.144

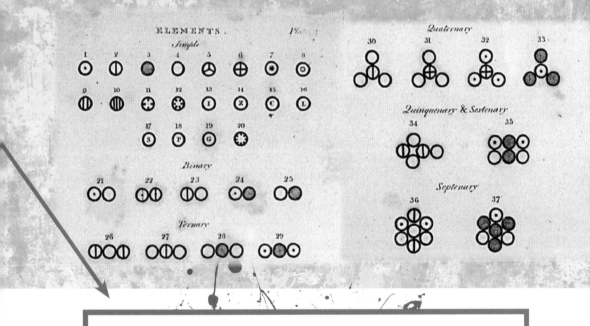

3/ Hack: Great thinkers have been speculating on the existence of atoms for thousands of years, but Dalton's atomic theory stands out as important.

It suggested that the only way to explain the behaviour of chemical elements is by accepting they are composed of discrete particles, or atoms.

No.70
The kinetic theory
of gases Are gases atomic?

1/Helicopter view: The scientists of the 19th century could not agree on the existence or otherwise of atoms (see: **Dalton's atomic theory**, page 142). This did not stop some of them explaining the behaviour of matter in a way that assumed the atomic world was real. The **kinetic theory of gases** certainly assumes the existence of atoms.

In its modern form, the kinetic theory of gases arguably began with **Daniel Bernoulli** in the 1730s. Bernoulli was working a few decades after scientists had begun to explore the relationship between the volume a gas occupies and its pressure and temperature. **Robert Boyle** had suggested these relationships could be understood by imagining gases to be composed of tiny particles that repel one another like springs. Many assumed that these particles moved little, if at all, and simply repelled each other from a distance.

Bernoulli questioned this assumption, suggesting instead that the particles in a gas are in very rapid motion and constantly colliding with the walls of the vessel in which the gas is confined – providing a *dynamic* explanation for pressure. The idea didn't catch on.

Its fortunes changed in the second half of the 19th century, when the scientific consensus on the nature of energy shifted (see: **The first law of thermodynamics**, page 128). Physicists including **James Clerk Maxwell** and **Ludwig Boltzmann** began to take the kinetic theory more seriously. They explored the mathematics and statistics of how tiny, independently moving particles in a gas should move and collide with one another, redistributing heat as they did so. This work confirmed that the behaviour of gases could indeed be explained by assuming they were composed of unimaginably small atoms or molecules moving at speed.

Even so, sceptics were quick to point out that the kinetic theory of gases did not *prove* the existence of atoms.

Daniel Bernoulli // 1700–1782

The atoms in gases move and collide at speed, just as popping corn kernels in a pan do.

2/Shortcut: Think of the particles in the kinetic theory of gases as corn kernels popping in a popcorn pan. If the pan is very large, at any given time perhaps only one or two kernels will collide with the lid as they pop, and the lid will remain firmly in place. However, if the pan is much smaller – but still contains the same number of kernels – dozens of them will simultaneously collide with the lid as they pop. The "pressure" caused by these popcorn "particles" is much higher. The pan's lid might move or even fall off as a consequence of all the collisions.

See also //
61 Boyle's Law, p.126

3/Hack: The behaviour of fluids, and the heat they carry, can be explained by assuming they are composed of atoms moving at great speed.

But the kinetic theory of gases does not *prove* the existence of atoms.

145

No.71
Einstein's theory of Brownian motion
Atoms revealed

 1/ Helicopter view: In 1827, **Robert Brown** set out to investigate the role of pollen in fertilization. He placed pollen in water and examined it under a microscope. Brown noted something strange: the pollen released tiny particles, and these particles "danced" in the water. About eighty years later, this dance – Brownian motion – would help convince many scientists of the existence of atoms.

Brown could not explain the movement of the tiny particles, but in the early 20th century, **Albert Einstein** found that he could. At the time, Einstein was trying to calculate the theoretical size of sugar molecules. One way to do this was to explore how sugar molecules diffuse (spread) through water to create a sweet solution. Einstein came up with equations to explain exactly how the sugar molecules should move and interact with the water molecules.

By 1905, Einstein realized his equations should explain how *other* molecules or particles interact with water molecules – even particles large enough to be visible through a microscope. The jittery dance of Brownian motion was just the sort of movement that Einstein's equations predicted would happen if tiny particles from pollen were under constant bombardment from energetic water molecules. The idea that the curious dance could be explained this way became known as **Einstein's theory of Brownian motion**.

Others had already suggested that the behaviour of gases or liquids could be explained by imagining they were composed of tiny, moving particles (see: **The kinetic theory of gases**, page 144) – but many scientists argued that these particles were purely theoretical. However, under Einstein's theory of Brownian motion, the strange dance Brown's grains performed was a *predicted consequence* of assuming that gases and liquids are made of moving atoms. Crucially, rival ideas did *not* predict the complicated dance of Brownian motion. The tide had turned. Most scientists were prepared to accept the existence of atoms.

The movement of dye through liquid – an example of Brownian motion – is explained by mathematics.

2/ Shortcut: Brown was not the first scientist to observe Brownian motion, but it was perhaps because he examined the phenomenon so thoroughly that it now carries his name. He first assumed that the erratic dance must relate to the vitality of living matter. He tested this idea with particles progressively further away from the "living" realm: pollen from recently dead plants, particles from fossil plants dating back two hundred million years, and eventually even particles from completely inorganic sources. No matter what the source, the particles always jumped around in water in exactly the same way. Brownian motion was not a property of living things – it had a deeper cause.

See also //

69 Dalton's atomic theory, p.142

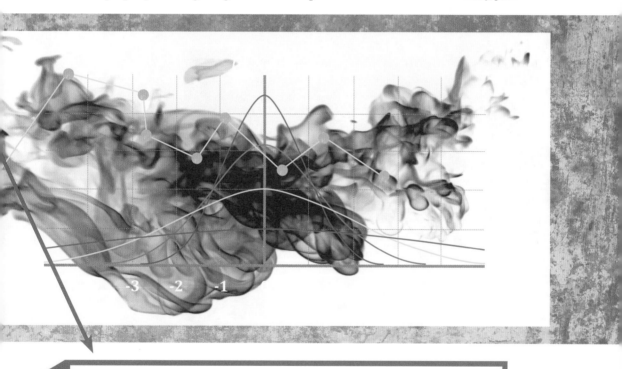

3/ Hack: Einstein's theory of Brownian motion may have settled an eighty-year-old mystery, but it also did something more profound.

It convinced many sceptics about the existence of atoms.

No.72
The plum pudding model
Atoms on the dessert menu

 1/ Helicopter view: Physicists of the early 20th century gradually came to accept the existence of atoms (see: **Einstein's theory of Brownian motion**, page 146). But doing so opened up a deeper mystery. What exactly do atoms look like?

Clues to the atom's structure had begun to emerge in the 19th century, even before the concept of atoms became widely accepted. Physicists including **Johann Hittorf** and **William Crookes** had discovered that a positively charged chunk of metal (a "cathode") releases a strange glow if it is held in an airless container. Further experiments revealed that these "cathode rays" carried energy and – given that the rays could be deflected by a magnetic field – that they were negatively charged.

By 1897, **Joseph John Thomson** had discovered that all sorts of materials produce cathode rays. From the way these rays behaved, they seemed to carry very tiny quantities of mass. This suggested the rays were actually streams of tiny particles. Thomson called them "corpuscles". Most other physicists opted to call them "electrons".

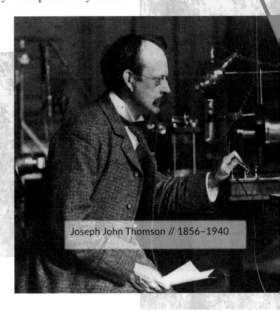

Joseph John Thomson // 1856–1940

Within a few years, research by **Henri Becquerel** and **Ernest Rutherford** had indicated that these electrons actually came from *inside* atoms. This breakthrough gave physicists an early clue about the internal atomic structure.

Thomson began thinking about that atomic structure. Atoms seemed to have no overall electric charge, so they had to contain a positively charged component to balance the negative charge of the electrons. The model he came up with assumed the atom was essentially a positively charged sphere with tiny negatively charged electrons embedded within it. The model looked a little like a fruitcake, with the dried fruit as electrons. As a consequence, Thomson's atomic structure became known as the **plum pudding model**.

An early model of the atom was reminiscent of a stodgy dessert.

2/ Shortcut: The name "plum pudding model" sounds flippant, which fits with the fact that Thomson's atomic vision was quickly abandoned. But his model was based on reason and logic. He drew on the work of **Alfred Marshall Mayer**, who had shown that dozens of tiny magnets arrange themselves in regular concentric rings when they are allowed to float freely in a water bath held beneath a strong magnet. If electrons inside atoms did the same, the patterns these rings formed could explain the macroscopic properties of chemical elements: two or more elements with similar physical properties would be built from atoms that contained similar patterns of electron rings.

See also //

75 The Rutherford–Bohr model, p.154

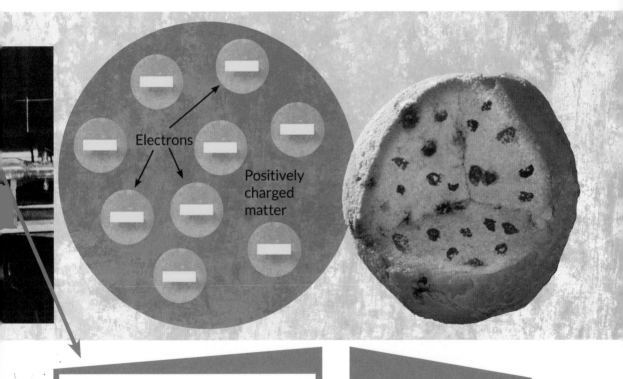

Electrons

Positively charged matter

3/ Hack: By the beginning of the 20th century, physicists wanted to understand the internal workings of the atom.

The plum pudding model was an early, although ultimately unsuccessful, vision of atomic structure.

No.73
Planck's law The dawn of a quantum revolution

Max Planck // 1858–1947

1/ Helicopter view: In the 1850s and 1860s, **Gustav Kirchhoff** argued that an object held at a constant temperature must emit exactly as much radiation as it absorbs in order to obey what became the **second law of thermodynamics** (page 130). It was an idea that would ultimately kick-start an entirely new field: quantum physics.

From his idea, Kirchhoff reasoned that a perfect radiation absorber must also be a perfect emitter of radiation. He called this perfect emitter a "black body" and said it was key to understanding radiation. Black objects behave a little like black bodies. For instance, on a sunny day a black surface is hot to touch because it absorbs sunlight (electromagnetic radiation) and emits thermal radiation.

Unfortunately, *perfect* black bodies are extraordinarily rare. Kirchhoff improvised. He suggested poking a tiny hole in the side of a large box. Because any light that enters the box through the hole is unlikely to escape again, the hole is an almost perfect radiation absorber: an almost ideal black body. If the sealed box is actually a hot oven, some thermal radiation will escape through the tiny hole – and it will be radiated almost "perfectly".

By the mid 1890s, **Wilhelm Wien** and **Otto Lummer** had put Kirchhoff's idea into practice and measured the thermal radiation that a hole in an oven emits at different temperatures. Physicists tried to find an equation that could explain the measurements – and they struggled.

In 1900, **Max Planck** solved the problem. But to do so he had to assume that the radiation emitted through the hole came in discrete chunks. Planck was apparently unhappy about having to factor in this assumption. But he had actually stumbled upon the quantum nature of the subatomic world. His findings became known as **Planck's law**.

Explaining the radiation from a "perfect" black body led to quantum physics.

2/ Shortcut: Physicists struggled to devise an equation that perfectly predicted the radiation emitted from Kirchhoff's "black body" at all temperatures. In what he described as an "act of desperation", Max Planck worked in a caveat that energy had to have discrete, whole number values – the equivalent of 1, 2, 3, and so on – rather than varying on an infinitely gradational scale. Planck named these discrete packets of energy "quanta". His work was the first foray into a new realm: quantum physics.

See also //

74 The photoelectric effect, p.152

75 The Rutherford–Bohr model, p.154

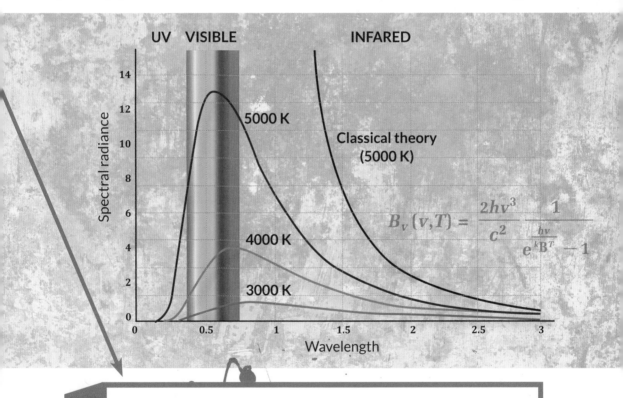

UV   VISIBLE   INFARED

Spectral radiance

5000 K

Classical theory (5000 K)

4000 K

3000 K

Wavelength

$$B_v\,(v,T) = \frac{2hv^3}{c^2}\; \frac{1}{e^{\frac{hv}{k_BT}} - 1}$$

3/ Hack: Planck's law perfectly describes the nature of the electromagnetic radiation emitted by a black body when it is in thermal equilibrium at a given temperature.

But it works only if it is assumed that the radiation is carried in discrete chunks, or quanta.

No.74
The photoelectric effect
Quantum theory gains momentum

 1/Helicopter view: In 1900, **Max Planck** found that he could account for the behaviour of electromagnetic radiation only by assuming the radiation came in the form of discrete packets, or quanta. Few people, including Planck, realized how important the discovery was – until **Albert Einstein** came along.

Heinrich Hertz had, in 1887, performed some experiments into light's ability to influence the electrical behaviour of certain metals. He was trying to create a tiny electric spark between two metal electrodes – something that would cause a small electric current to flow. Mysteriously, this current became more likely to flow if the metals were bathed in ultraviolet light.

Hertz left this mystery unsolved. But over the next 15 years many others explored this **"photoelectric effect"**. It soon became clear that the electric spark between the electrodes in Hertz's experiment was actually a flow of tiny subatomic particles dubbed "electrons" (see: **The plum pudding model**, page 148).

At the turn of the 20th century, **Philipp Lenard** made an important discovery. By now it was reasonably clear that the light bathing the experiment must be adding energy, and that this energized electrons in the metal so much that they jumped off its surface to form the electric spark. But Lenard discovered that the energy those electrons carried related only to the *colour* of the light not its *brightness*. This was not how light waves should behave.

Heinrich Hertz // 1857–1894

In 1905, Einstein suggested a solution. He argued that light must come in the form of discrete particles, or packets – compatible with Planck's quanta – each of which carried a certain amount of energy. Exactly how much energy they carried depended on the colour of light, not its intensity. Einstein's solution was not popular. **Robert Millikan** spent several years trying to disprove it through experiments. But he could not. The quantum theory was gaining ground.

Each colour of the rainbow has a different effect on the electrons in some metals.

2/ Shortcut: Physicists at the turn of the 20th century thought light behaved as a wave. If the energy in these waves knocked electrons off the surface of a metal, it stood to reason that a bigger wave should have a greater effect. But making light waves bigger (making the light brighter) did *not* generate more of an electric spark. What *did* generate a larger spark was making the waves *shorter* (changing the light's colour from red to violet). Einstein argued that light must actually come in discrete particles, with "red" particles carrying less energy than "violet" ones. Because "red" particles are weak, bombarding metal with millions of them would have a small impact. But bombarding the metal with even a handful of strong "violet" particles could have a big impact.

See also //

73 Planck's Law, p.150

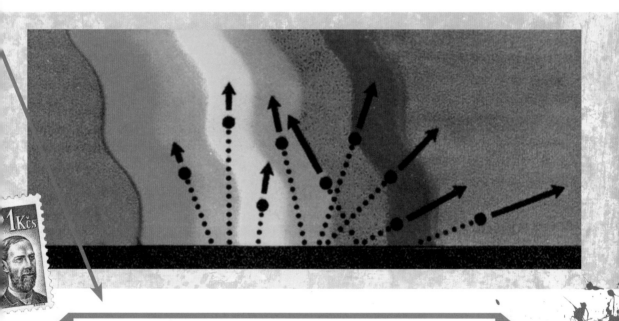

3/ Hack: The photoelectric effect was one of the most puzzling discoveries of the late 19th century. Einstein's explanation for it had two implications.

It revived the particle theory of light and it bolstered the quantum theory of subatomic physics.

No.75
The Rutherford–Bohr
model Atoms as planetary systems

 1/ Helicopter view: Early in the 20th century, **Ernest Rutherford** and his colleagues **Hans Geiger** and **Ernest Marsden** made an important discovery about the behaviour of atoms – one that hinted that atoms are mostly empty.

Rutherford had already gained a reputation for his work with radiation. Exploring further, he and his colleagues discovered that one type of radioactive particle – the "alpha particle" – typically sails through an extraordinarily thin sheet of gold. Very rarely, however, the gold sheet significantly deflected an alpha particle – a result as surprising as discovering that a sheet of tissue paper could deflect a missile.

In 1911, Rutherford suggested that this result hinted at the internal structure of the atom. He argued for a Solar System-like atom. At its centre was a tiny, dense, charged area – later named the atomic nucleus – which was surrounded by a larger cloud of low-density electrons. Most alpha particles sailed through the gold sheet because they passed through the electron cloud. Once in a while, though, an alpha particle collided with the dense nucleus and simply bounced off.

A few years later, **Niels Bohr** began working with Rutherford, and the pair refined this model. Bohr was convinced that energy comes in discrete packets or "quanta" (see: **Planck's law**, page 150). He suggested that electrons occupy *discrete orbits* around the atomic nucleus. He said they could jump between orbits by either absorbing or emitting packets of energy.

Bohr's 1913 atomic model received a boost the following year when experiments by **James Franck** and **Gustav Ludwig Hertz** seemed to confirm the quantum behaviour of atomic electrons. With time, physicists would further revise their understanding of atomic structure, but the atomic model school children learn about even today is based on the **Rutherford–Bohr model**.

Hans Geiger //
1882–1945

Ernest Marsden //
1889–1970

Electrons orbit a central nucleus in the Rutherford-Bohr atomic model.

2/Shortcut: The Rutherford–Bohr model predicted that an electron in an atom would jump between "orbits" by absorbing a discrete packet of energy. Franck and Hertz's 1914 experiment was seen by many to confirm this prediction. Franck and Hertz accelerated electrons through a vapour composed of mercury atoms. The electrons gradually gained speed (and energy) as they moved through the vapour, but typically they suddenly lost all of this energy when it had built up to a specific level. The explanation was that each free-moving electron eventually gained enough energy to kick an electron inside a mercury atom up to a different "orbit". At this point, one of those atomic electrons effectively "stole" all of the free-moving electron's energy, instantly decelerating it.

See also //

82 The concept of nuclear transmutation, p.168

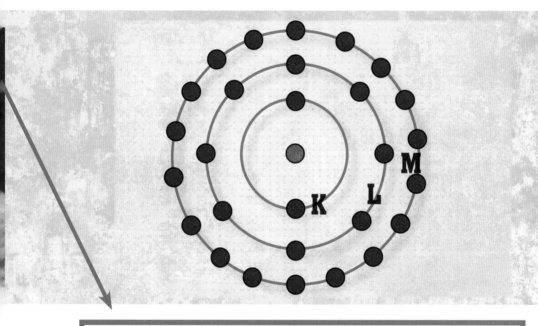

3/Hack: The Rutherford–Bohr model combined ideas from traditional (classical) physics with concepts from the new field of quantum physics.

It imagines atoms as tiny Solar Systems, with electron "planets" traveling around an atomic nucleus "Sun" in discrete orbits.

No.76
Heisenberg's uncertainty principle How to be certain what it means

1/Helicopter view: Quantum theory seemed to be in trouble by the early 1920s. Atoms were viewed like tiny Solar Systems, with electrons orbiting a central nucleus (see: **The Rutherford–Bohr model**, page 154). But although this model worked well for hydrogen atoms – with just one orbiting electron – it was less successful when applied to complicated atoms with multiple electrons. **Werner Heisenberg** came up with a drastic solution: abandon the idea that electrons orbit the nucleus in a classical sense.

Heisenberg insisted that physicists should accept that they could never actually *observe* fundamentals like the position and momentum of electrons in atoms. It was simply impossible to treat electrons as tiny "planets" moving in a steady and predictable way inside atoms.

This was not a popular idea. Other physicists, including **Albert Einstein**, pointed out that there were situations in which electrons *did* move in steady and predictable ways. For instance, in some experiments, electrons were fired through chambers filled with condensed gas – and they left little trails in their wake. The electrons had taken a steady and (potentially) predictable path. This implied that electrons also moved in predictable ways inside atoms.

Eventually Heisenberg argued that these "cloud chamber" experiments only gave the *impression* that electrons had taken a predictable path through the cloud. Strictly speaking, what physicists were really seeing was a series of interactions between an impossibly small electron and a number of much larger droplets in the cloud. Collectively, these interactions seemed to imply the electron had travelled in a nice and steady way through the cloud. But in reality, argued Heisenberg, each interaction gave only a very *approximate* sense of where the electron was and its momentum at any given point in time. Heisenberg said this vagueness was an inherent and fundamental property of the subatomic world. It became known as **Heisenberg's uncertainty principle**.

Electrons traveling through a cloud chamber leave tiny trails in their wake.

2/ Shortcut: In discussing his uncertainty principle, Heisenberg often cited a conversation he once had with **Burkhard Drude**. Drude argued that a sufficiently powerful microscope should be able to resolve *exactly* where an electron was inside an atom. Heisenberg realized that such a microscope would have to use light of an incredibly short wavelength in order to interact with something as tiny as an electron. But as wavelength becomes shorter, light also becomes more energetic. Light with a sufficiently short wavelength to "see" an electron would carry so much energy, it would inevitably change the electron's momentum in an unpredictable way. Even if Drude's microscope could tell us exactly where an electron is at a given point in time, it could not reveal its exact momentum.

See also //

77 The complementarity principle, p.158

78 The EPR paradox, p.160

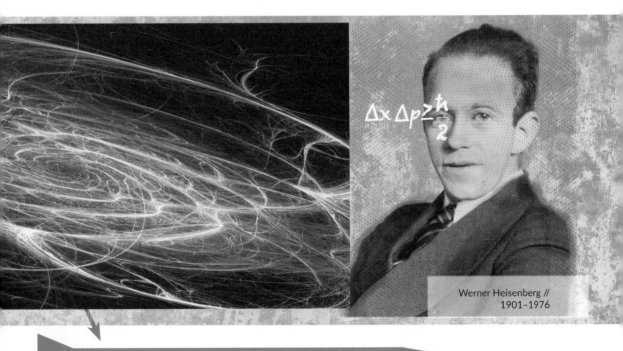

$$\Delta x\, \Delta p \geq \frac{h}{2}$$

Werner Heisenberg //
1901–1976

3/ Hack: Heisenberg's uncertainty principle puts a fundamental limit on our ability to define complementary features of subatomic particles, such as their position and momentum.

It is an idea at the very core of quantum physics.

No.77
The complementarity
principle When words lose
their power

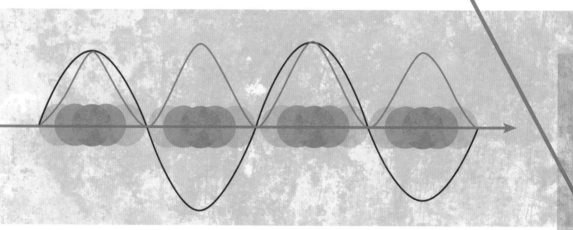

 1/ Helicopter view: Niels Bohr did a great deal of thinking about the subatomic world in the 1920s. He was well aware (as were other physicists) that it was dogged by a series of fundamental, but probably interconnected, contradictions.

The most notorious of these contradictions concerned light. Over the centuries, opinion had swayed first this way and then that on whether light was best described as a series of discrete particles or as waves traveling through some sort of medium.

At the turn of the 18th century, **Isaac Newton** seemed to have settled the debate in favour of particles (see: **The corpuscular theory of light**, page 120). But a century later, **Thomas Young** convinced many that light is really a wave (see: **The wave theory of light**, page 122). In the late 19th century, **James Clerk Maxwell** bolstered this wave theory (see: **Maxwell's equations**, page 124), only for **Albert Einstein** to provide convincing evidence in favour of the particle theory a few decades later (see: **The photoelectric effect**, page 152).

What did it all mean? In 1927, Bohr decided on an astonishing compromise. The experimental evidence on both sides was so convincing, he said, that physicists simply had to accept that light was *both* a particle *and* a wave. Or, at least, it sometimes behaved in a way we would describe as wave-like and sometimes behaved in a particle-like manner. This isn't a contradiction, insisted Bohr, but simply two *complementary* pictures of nature. Bohr's idea, which is known as the **complementarity principle**, would become central to the science of quantum physics.

2/Shortcut: The complementarity principle is partly a philosophical argument. Bohr was convinced that, at the subatomic level, nature behaves in a way so alien to our expectations that it is impossible for us to really picture it properly. Worse than that, because even our language is based on our expectations of a world that obeys "classical" physics, we can't even *describe* the subatomic world adequately. This is why light sometimes seems to behave in a way we would define as wave-like, while at other times it definitely seems to behave as a particle. The problem is not with light, it is with the language we use to describe it.

The quantum world is difficult to describe with words.

See also //

78 The EPR paradox, p.160

80 The electron double-slit experiment, p.164

Niels Bohr // 1885–1962

3/Hack: Scientists had been arguing for centuries about the true nature of subatomic phenomena like light. The complementarity principle explains why the debate lasted so long.

Words lose their power to fully explain what is going on at the subatomic level.

No.78

The EPR paradox
Einstein's attack on quantum physics

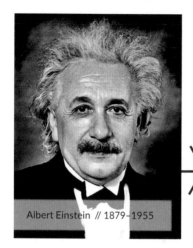

Albert Einstein // 1879-1955

1/ Helicopter view: Niels Bohr and Werner Heisenberg began arguing, in the late 1920s, that there is inherent uncertainty at the subatomic level that no amount of careful scientific investigation could remove (see: **Heisenberg's uncertainty principle**, page 156). **Albert Einstein** strongly rejected this claim. He felt that built-in uncertainty was evidence that the emerging quantum theory had to be incomplete – God, he said, "does not throw dice". The **EPR paradox** of 1935 was one of his most famous attempts to make his point.

The EPR paradox is so named because Einstein published it in collaboration with **Boris Podolsky** and **Nathan Rosen** (EPR being their combined initials). It was a thought experiment designed to show the imperfections in quantum theory.

The three researchers realized that the emerging quantum theory predicted that two subatomic particles could interact in such a way that their properties become intimately linked – and that they would remain linked even if they then drifted away from one another. **Erwin Schrödinger** would later term this curious phenomenon "quantum entanglement".

Einstein and his colleagues argued that entanglement should in principle allow an observer to accurately measure *both* the position *and* the momentum of a particle – something that Heisenberg's uncertainty principle stated was impossible.

Unfortunately for Einstein and his colleagues, the legacy of the EPR paradox hasn't played out in the way they had hoped. Many physicists actually began exploring the idea of entanglement and confirmed that it does indeed occur. Today, the phenomenon underlies many emerging quantum technologies. Far from demonstrating that quantum theory is incomplete, the EPR paradox is seen to have helped show just how fundamentally strange the quantum world really is.

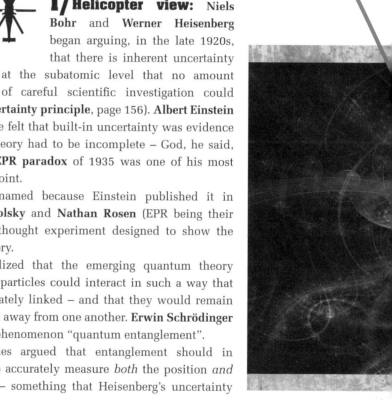

Einstein used two particles linked by quantum entanglement to attack quantum theory.

2/Shortcut: There are a few ways to interpret the EPR paradox – one is to imagine a subatomic particle spontaneously splits into two – A and B – that whizz off in opposite directions. The physical properties of A and B are intimately related. In a sense they are like mirror twins. If physicists measure the exact location of A, logically they must know the exact location of B even without looking at it. If they then measure the exact momentum of A, again logic dictates that they know the exact momentum of B. This means the researchers have now established both B's precise location *and* its precise momentum. Using quantum entanglement they have apparently disproved Heisenberg's uncertainty principle.

See also //

77 The complementarity principle, p.158

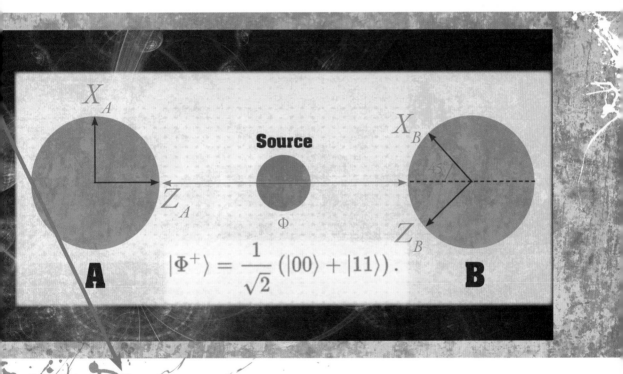

$$|\Phi^+\rangle = \frac{1}{\sqrt{2}}\left(|00\rangle + |11\rangle\right).$$

3/Hack: The EPR paradox was one of Einstein's most famous attempts to attack the emerging quantum theory.

But many physicists think the paradox actually helps demonstrate how strange the quantum world really is.

No.79
Schrödinger's cat
Quantum physics reaches peak weirdness

1/Helicopter view: In the 1930s, **Erwin Schrödinger** became deeply uneasy about the implications of quantum theory. When **Albert Einstein** and his colleagues introduced a paradox to try to show that the theory was incomplete (see: **The EPR paradox**, page 160), Schrödinger became even more unsettled. His discomfort led him to introduce the world to a now-famous cat.

Schrödinger gave a name to the strange phenomenon the EPR paradox exploited: quantum entanglement. When he combined this with another strange predicted consequence of quantum physics called "quantum superposition", Schrödinger reached philosophically troubling conclusions.

Quantum superposition states that the physical properties of a subatomic particle are not completely determined until the moment it is actually measured. For instance, physicists describe one fundamental subatomic property using the term "spin". A particle may spin "up" or "down". But until the particle is actually measured, quantum theory insisted it spins *both* up *and* down.

What worried Schrödinger was that this quantum superposition could, in principle, become part of the entanglement between two subatomic particles. Two entangled particles – A and B – could end up sharing the same quantum superposition. Particle A would spin both up and down, and particle B would also spin both up and down. Only once one of the two particles was measured would the fates of both be sealed: if measurement confirmed that particle A spins "up", particle B – its entangled mirror twin – must spin "down".

Schrödinger was concerned that the quantum superposition could, in principle, "entangle" anything with which it interacted. In 1935, he pointed out that this could include objects much larger than subatomic particles – a cat, for instance. Our experience tells us cats and other animals are either alive or dead. But if a feline became mixed up in an entangled quantum superposition, Schrödinger realized it could, paradoxically, be *both at the same time*.

Schrödinger's cat is a deliberately absurd idea.

2/Shortcut: Schrödinger imagined a steel chamber containing a cat and a tiny fragment of a radioactive substance. Over the course of an hour, an atom in the substance might (or might not) spontaneously shoot out a radioactive particle. Until physicists open the box to check, the atom is in a quantum superposition: it both *has* and has *not* released a radioactive particle. If it has released the particle, though, Schrödinger's thought experiment imagined this particle triggering a chain reaction that ultimately poisons – and kills – the cat. This means the cat itself has become entangled in the quantum superposition. It is both alive *and* dead.

See also //

98 The many-worlds interpretation, p.200

$$|b\rangle = \frac{1}{\sqrt{2}}\left(|00\ldots0\rangle + |11\ldots1\rangle\right)$$

Erwin Schrödinger //
1887–1961

3/Hack: Schrödinger's cat has become one of the most famous thought experiments in quantum physics because it seems to jar against our expectations of reality.

This was exactly the point Schrödinger wanted to make.

No.80
The electron double-slit
experiment
Demonstrating quantum weirdness

Louis De Broglie // 1892–1987

1/ Helicopter view: Many physicists of the late 1930s were prepared to accept quantum theory, even though it seemed to imply that the world was capable of behaving in a nonsensical way (see: **Schrödinger's cat**, page 162). There is a simple reason why the theory became popular. For all its apparent paradoxes, quantum theory correctly predicted the outcomes of experiments. One of the best examples is provided by the **electron double-slit experiment**.

In the early 19th century, **Thomas Young** had shown convincingly that light mixes and interacts when it passes through two slits, just like water waves might be expected to do, implying light was a wave. But with the advent of quantum theory, physicists began to accept that light actually behaves as *both* a wave *and* a particle (later dubbed the photon).

In the early 1920s, **Louis De Broglie** argued that this "wave-particle duality" held true for *all* particles – and most physicists agreed with his assessment. This gave physicists an opportunity to revisit Young's double-slit experiment, this time using electrons rather than light.

Claus Jönsson and his colleagues were the first to do so in the early 1960s. They found that electrons produced a wave-like interference pattern as they passed through two slits just as Young had found that light does. About a decade later, in 1974, **Pier Giorgio Merli** and his colleagues repeated the experiment – but using a setup that allowed them to fire electrons one at a time at the two slits. The individual electrons behaved in a way that makes sense only if *each* is behaving as both a wave and a particle. Merli's team had admirably demonstrated that the strange wave-particle duality predicted by quantum physics could be confirmed through experiments.

A single electron "particle" can also behave as a wave that simultaneously "washes" through two slits.

2/Shortcut: Merli and his colleagues fired electrons at the double-slit screen one at a time. Intuitively, each electron *particle* should sail through one of the slits and hit the detector behind – which is what the team found. But after the scientists had fired tens of thousands of electrons (individually) at the double-slit, the tiny dots on the screen built up into a series of bars – a wave-like interference pattern. They concluded that each electron "particle" had behaved as a relatively broad *wave* that "washed" through both slits at the same time. On the far side of the two slits, a given electron's two sets of "ripples" interfered as water waves might. But when these ripples reached the detector, they collapsed into a single particle-like dot, the location of which was determined by the wave interference pattern.

See also //
59 The wave theory of light, p.122

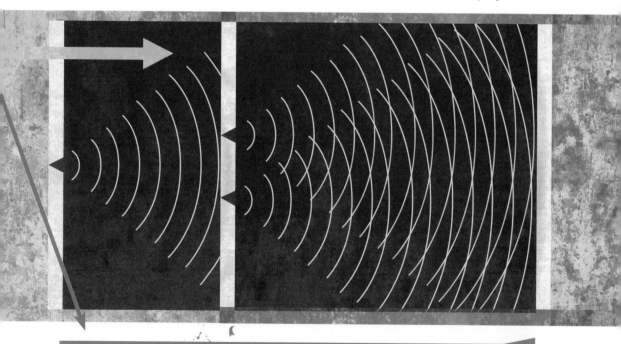

3/Hack: Quantum physics makes bold, almost unbelievable predictions. But the electron double-slit experiment confirms some of those predictions.

And for physicists, seeing is believing.

No.81
The concept of antimatter
Why does it matter?

1/ Helicopter view: **Paul Dirac** was interested in the way that the quantum world should behave if **Albert Einstein**'s theories of relativity were correct. In the 1920s, he realized that marrying the two ideas led to a strange prediction: there should be a suite of never-before-seen "antiparticles" – the antimatter version of subatomic particles.

Within a few years, Dirac's prediction proved correct. **Carl Anderson** was experimenting with electrons – subatomic particles with a negative charge He discovered that some of them behaved in a way that suggested they were actually *positively* charged. They were "anti-electrons". Anderson named them positrons.

Despite the exotic name, antimatter is relatively easy to create in some of today's physics laboratories. For instance, whenever particles slam into each other in the Large Hadron Collider on the Swiss-French border, the debris contains new particles of antimatter (and matter). The antimatter does not survive very long, though, because as soon as matter and antimatter come into contact with one another, they annihilate, leaving nothing but radiation behind.

This destructive process leaves physicists with a puzzle. Matter and antimatter should have been created in equal quantities at the birth of the Universe (see: **The big bang theory**, page 188) and so should have destroyed each other completely in subsequent matter-antimatter collisions. This means that the observable Universe should, in theory, contain *neither matter nor antimatter*. As far as scientists can tell, there is, indeed, very little antimatter out there.

But stars and planets (and humans) are clearly built from matter. This suggests that, although matter and antimatter were created in equal quantities at the dawn of the Universe, the antimatter later behaved in a way that made its destruction *more likely* than the destruction of matter. Scientists are still trying to explain how and why this happened.

Paul Dirac's predicted existence of antimatter is confirmed at experiments like the Large Hadron Collider.

2/ Shortcut: Imagine growing up believing you had no siblings only to be told that you were actually separated at birth from a twin. The discovery would almost certainly trigger all sorts of questions about this mysterious "other self". Physicists faced a very similar situation in the first half of the 20th century. They had been aware of the existence of matter for centuries. Suddenly they were confronted with the idea that antimatter existed too, and their world view changed forever.

See also //

72 The plum pudding model, p.148

Paul Dirac // 1902–1984

3/ Hack: Antimatter behaves in a broadly similar way to familiar matter.

The only significant differences concern properties including the electric charge on the particles.

No.82
The concept of nuclear transmutation
Real world alchemy

Ernest Rutherford // 1871–1937

1/ Helicopter view: By the end of the 19th century, alchemy had long been considered flawed science. No scientist seriously accepted the existence of a substance – the philosopher's stone – with the power to transform, or "transmute", common metals like lead into precious gold or silver. Then, in the early 20th century, this view began to change.

Two physicists – **Ernest Rutherford** and **Frederick Soddy** – may have been the first to realize that transmutation is possible after all. In 1901, the pair discovered that thorium, a radioactive chemical element, was naturally transforming into a *different* element called radium as it radioactively "decayed".

Within a few decades, Rutherford and other physicists were able to achieve nuclear transmutation artificially: they found that by firing subatomic particles at an atom, they could chip off chunks from the atom's nucleus, changing its configuration and transmuting it into another element.

What of turning lead into gold? At the atomic level these two elements are rather similar: the key differences are in the atomic nucleus. Lead has 82 protons in its nucleus while gold has just 79. To transmute lead into gold, all that is required is to fire subatomic particles at lead atoms until three protons are knocked out of the nucleus.

Reportedly, the feat was first achieved by accident in the 1970s at a Soviet nuclear facility. When the researchers examined the lead shielding used in their reactor, they discovered that some of the lead atoms had turned to gold. In a sense, today's nuclear physicists are alchemists – but few would be willing to use the term.

Lead can become gold, but it's not an easy change to make.

2/Shortcut: Chemical elements, including gold, are defined by the structure of their atoms. More specifically, it is the number of protons – positively charged subatomic particles – in the atomic nucleus that determines the identity of a chemical element. Alchemists believed they could transmute one element into another using what we would recognize today as chemical reactions. This is impossible. But modern physics *does* offer a way to change the number of protons in the nucleus, and hence transmute one element into another.

See also //
83 The concept of radiometric dating, p.170
84 The nuclear electron hypothesis, p.172

82 Pb 207.2
2 8 18 32 18 4

79 Au 196.9665
2 8 18 32 18 1

3/Hack: Contrary to popular thought, it is possible to transform lead into gold. But the process requires a lot of expensive equipment and specialist knowledge.

It is an impractical way of making a fortune.

No.83
The concept of radiometric dating
The age of the Earth revealed

1/ Helicopter view: In the 1860s, **William Thomson** (later **Lord Kelvin**) applied his understanding of thermodynamics to a long-standing question: how old is the Earth?

Over the next few decades he argued that the temperature of the Earth suggested our planet could be no more than a few tens of millions of years old. Thomson's conclusion brought him into intellectual conflict with geologists and biologists, who thought that the Earth was much older (see: **The concept of deep time**, page 74). Thomson found their ideas vague and qualitative. If only there was a *quantitative* way to measure the Earth's age. At the turn of the 20th century, such a method began to emerge.

The story begins in the mid-1890s with **Henri Becquerel**, who discovered that uranium-containing minerals spontaneously produced a mysterious form of radiation. **Pierre** and **Marie Curie** explored the phenomenon in more detail and gave it a name: radioactivity.

In the early years of the 20th century, **Ernest Rutherford** and **Frederick Soddy** realized that radioactive elements like uranium actually "decay" – change their identity to become new elements – as they release radiation. By 1904, Rutherford had begun arguing that the phenomenon could act as a "clock" for dating rocks.

Rutherford's dating concept was crude, but his research would lead to better, more accurate versions. He noted that there was an inherent randomness to the generation of radioactivity – an early sign of the inherent uncertainty in subatomic processes (see: **Heisenberg's uncertainty principle**, page 156). But, averaged out over time, a pattern emerges: for instance, he calculated that it should always take about 2,600 years for *half* of a quantity of radioactive radium to decay. This idea of the "half-life" allowed researchers who followed to develop the **concept of radiometric dating**, and eventually convince scientists that the Earth was, after all, billions of years old.

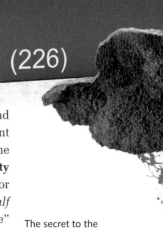

88
Ra
(226)

The secret to the age of Earth is hidden in rocks – and unlocked with nuclear physics.

2/ Shortcut: Rutherford was the first to attempt to use radiometric dating. He knew that uranium-containing minerals decay, releasing mysterious "alpha particles" in the process. He guessed (correctly) that those alpha particles would eventually be identified as a form of helium. This suggested that uranium-containing rocks should accumulate helium as they age. He measured how much helium there was in a rock sample in his possession, and – using an estimate for the rate of alpha particle (that is, helium) production – calculated it was forty million years old. It was a crude estimate, given that some of the helium might have leached out of the rock with time. Later scientists refined the technique.

See also //

82 The concept of nuclear transmutation, p.168

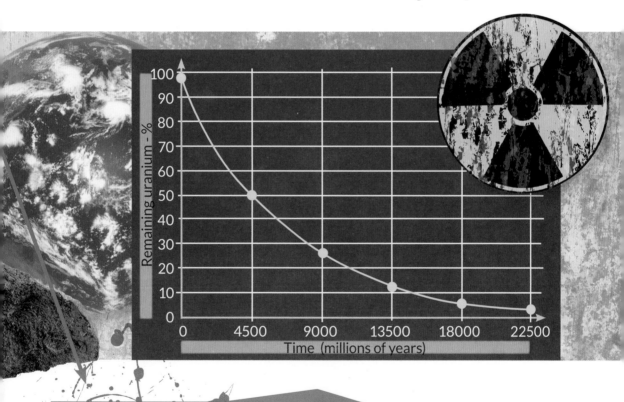

3/ Hack: For centuries, scientists argued about the true age of the Earth.

The concept of radiometric dating ultimately helped settle the debate and confirm our planet's antiquity.

No.84
The nuclear electron
hypothesis A puzzle at the core
of the atom

1/Helicopter view: In the mid-1910s, physicists developed a model of the atom in which electrons orbited a small, dense nucleus (see: **The Rutherford–Bohr model**, page 154). Many physicists focused their attention on understanding the electrons. But for others, the nucleus was more alluring. What was inside?

By 1919, **Ernest Rutherford** had made a breakthrough: he managed to knock off small positively charged particles from the nucleus of nitrogen atoms – particles that became known as protons.

The discovery was satisfying: the positive protons could balance the negative electrons to make atoms neutral. But there was a problem. Experiments suggested that the proton wasn't "heavy" enough to account for all of the mass inside the nucleus. There was something else in there too.

A leading hypothesis was that the nucleus must contain additional protons, and an equal number of electrons to balance their charge. This was an attractive idea: experiments had shown that the atomic nucleus could release radiation (dubbed "beta radiation") that appeared to consist of electrons. But the **nuclear electron hypothesis** was also beset by problems – not least the fact that **Heisenberg's uncertainty principle** seemed to imply that an unfeasibly large amount of energy would be needed to keep electrons confined in a space as small as the nucleus.

James Chadwick // 1891–1974

In 1932, **James Chadwick** rescued the situation by discovering another particle in the nucleus – one that was slightly heavier than the proton and, crucially, neutral in charge. The particle was dubbed the neutron. It solved the nucleus mass deficit, and implied the nucleus was built entirely of positive protons and neutral neutrons. But why did it sometimes spit out negative electrons? Chadwick's discovery left this puzzle unsolved, which might help explain why **Werner Heisenberg** continued to assume the nucleus must contain electrons for some time after the discovery of the neutron.

What exactly is going on inside the atomic nucleus?

2/ Shortcut: Chadwick's work ultimately helped disprove the popular nuclear electron hypothesis. He had long suspected the existence of a neutron-like particle in the atomic nucleus. In the early 1930s he heard about an experiment by **Frédéric** and **Irène Joliot-Curie**: they had bombarded a chemical element called beryllium with radiation and found that it kicked out its own radiation as a consequence. Chadwick realized that this beryllium radiation behaved in an odd way. It had no electric charge, but unlike known forms of "neutral" radiation, it could knock protons loose from atomic nuclei. He reasoned that, in order to have such an impact on "heavy" protons, the radiation itself must have a lot of mass. It had to be composed of neutrons.

See also //

76 Heisenberg's uncertainty principle, p156

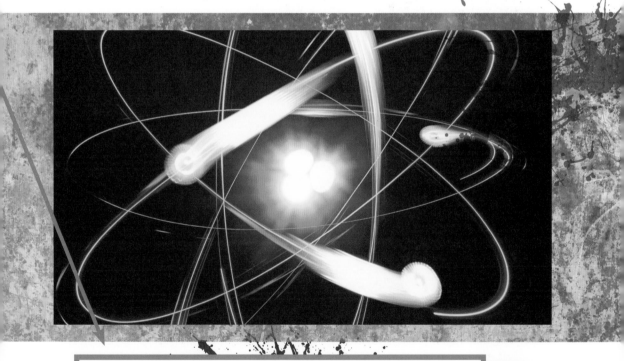

3/ Hack: The nuclear electron hypothesis was an early attempt to account for a mass deficit in the atomic nucleus, before the discovery of neutrons solved the problem.

But the hypothesis was important: it emphasized that the atomic nucleus could emit electrons, hinting that nuclear physicists of the 1920s and 1930s still had mysteries to solve.

No.85
The theory of nuclear
fission An idea with explosive implications

Lise Meitner // 1878–1968

1/Helicopter view: Early in the 1930s, **James Chadwick** discovered the neutron, a subatomic particle within the atomic nucleus. The discovery kicked off an intellectual chain reaction that would ultimately help end the Second World War in a devastatingly destructive fashion.

Within years of Chadwick's discovery, physicists realized the neutron was not only a missing piece in their understanding of the atomic nucleus – it was also a powerful tool for finding out even more. Earlier in the 20th century, physicists had discovered that they could chip chunks off the atomic nucleus by bombarding it with other subatomic particles. However, these subatomic particles were all positively charged – as was the nucleus itself. This limited the ability of the experiments to probe deeply inside the nucleus, because the incoming particles and the nucleus naturally repelled each other.

The neutron had no charge, and it was a relatively "heavy" particle – it could in principle penetrate far into the nucleus. By 1934, **Enrico Fermi** had begun bombarding uranium atoms with neutrons. In Berlin, **Lise Meitner**, **Otto Hahn** and **Fritz Strassmann** repeated some of Fermi's experiments and tried to identify the atomic products.

With the rise of the Nazis, Meitner – born into a Jewish family – fled Berlin to the safety of Sweden. Hahn and Strassmann continued working, and concluded that some of the uranium seemed to have become barium, an element with a much smaller nucleus. Meitner and her nephew, **Otto Frisch** – who had also fled to Sweden – realized that Hahn and Strassmann had "split the atom", dividing the uranium nucleus into two much smaller chunks and releasing energy. Frisch gave the process a name: nuclear fission. At the time, no one had thought fission was possible, but within a few years the **theory of nuclear fission** led to the creation of the atomic bomb.

2/Shortcut: Meitner and Frisch's nuclear fission reaction begins with a form of uranium containing 92 protons (and 143 neutrons) in its nucleus. Bombarding it with neutrons sometimes causes the nucleus to split in two, forming krypton (36 protons) and barium (56 protons). The combined mass of the barium and krypton is slightly less than the initial mass of the uranium – the extra mass is released as energy (see: $E = mc^2$, page 138). This energy kick is small. But if a neutron splits the atom in a way that releases two neutrons, those neutrons could split two further uranium atoms, releasing a total of four neutrons that could split four atoms, and so on. This rapid nuclear chain reaction could release lots of energy.

The atomic bomb – an end product of atomic investigations.

See also //

84 The nuclear electron hypothesis, p.172

3/Hack: The theory of nuclear fission suggested it was possible to divide large atomic nuclei into two or more much smaller fragments.

The process might have remained of purely academic interest but for the fact that fission can convert some mass into energy.

No.86
The quark model
Getting to the root of matter

1/Helicopter view: Physicists had identified dozens of distinct subatomic particles by the middle of the 20th century. A few physicists began to grumble about the difficulties of remembering all the members of this "particle zoo". Could it be simplified?

By the early 1960s, an attempt to do just that was beginning to crystallize. **Murray Gell-Mann** and **Yuval Ne'eman** independently came up with an idea that became known as the "eightfold way", which suggested that the particle zoo could be organized into a small number of families, each defined by a shared set of basic physical properties. This classification scheme worked well, but as even more particles were discovered, the eightfold way began to run into trouble.

In 1964, Gell-Mann (and, independently, **George Zweig**) suggested a radical extension of the eightfold way. Both physicists argued that many particles were actually composed of *even smaller* subunits. They suggested that a small number of truly indivisible or "elementary" particles could combine in different ways to create the diverse particle zoo.

Murray Gell-Mann // b. 1929

The new way of thinking had profound consequences. It suggested, for instance, that the protons and neutrons inside the atomic nucleus were each composed of a trio of elementary particles that Gell-Mann dubbed "quarks". Ultimately this would help explain the puzzle of why the nucleus sometimes kicks out an electron: a neutron can be converted into a proton by changing its quark configuration – and the process creates an electron (and also an antimatter particle called an electron antineutrino).

However, many physicists – apparently including Gell-Mann – were reluctant to accept that quarks really existed because physicists had never observed them. This situation began to change at the end of the 1960s, when experiments confirmed that protons did indeed seem to be composed of smaller subunits, just as Gell-Mann and Zweig predicted. The **quark model** moved toward general acceptance.

Physicists now say that the protons and neutrons inside the atomic nucleus each contain three quarks.

2/ Shortcut: The quark model proposed that many relatively large subatomic particles, including the proton and neutron, were actually built from two or three smaller quarks. By the late 1960s those quarks were beginning to reveal their presence. Physicists fired very small subatomic particles called electrons at much larger protons and carefully measured the way they behaved in an effort to learn more about the proton's structure. The experiment strongly implied that the proton is composed of smaller building blocks. Eventually, physicists agreed that these building blocks were Gell-Mann and Zweig's quarks.

See also //

87 The Standard Model, p.178

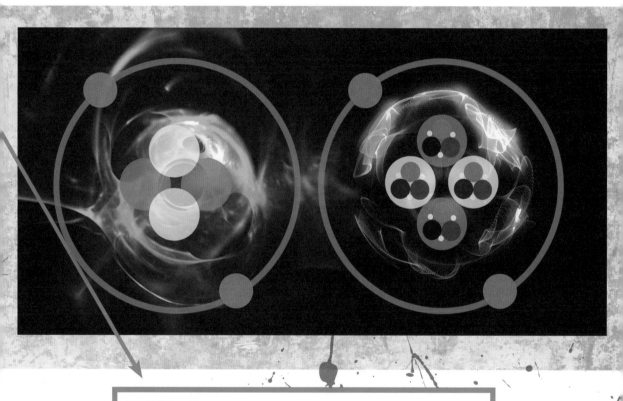

3/ Hack: By the mid-20th century, physicists had discovered an extraordinary diversity of subatomic particles. They suspected a simpler, underlying pattern must explain that diversity.

The quark model provides that pattern.

No.87
The Standard Model
A theory of (nearly) everything

1/Helicopter view: Quantum physics does more than explain the internal structure of atoms – it has come close to explaining the way the Universe functions at a fundamental level.

Many quantum physicists of the 1920s and 1930s were concerned with understanding the atom (see: **The Rutherford–Bohr model**, page 154). However, other researchers – particularly **Paul Dirac** – wanted to know whether quantum physics could explain the way that matter and energy *interact*, and so help explain the way the Universe *behaves*.

In the late 1920s, evidence began to emerge that it could. Quantum electrodynamics – "QED" – helped explain the interactions between light (electromagnetic radiation) and matter. QED claimed that subatomic particles called photons traveled at the speed of light between chunks of matter, like messengers carrying the electromagnetic force through spacetime. This idea brought together quantum physics and **Albert Einstein**'s **special theory of relativity**.

As physicists learned more about the subatomic world, they came to accept that there are a total of four fundamental forces permeating the Universe. They have now found good evidence that three of them – the electromagnetic, strong and weak forces – are carried between chunks of matter by a special class of subatomic particles termed bosons (the photon is a boson). There is even good evidence of a fourth type of boson – the Higgs boson – that interacts with matter to give it mass. The theoretical framework that incorporates these bosons and also the "matter particles" (see: **The quark model**, page 176) is known as the **Standard Model**. But the Standard Model is not complete. The fourth fundamental force – gravity – does not fit. In principle it, too, could be carried through spacetime via a boson. But physicists have found no solid evidence that this hypothetical boson (the "graviton") exists. Until they do, the Standard Model is only a theory of *nearly* everything.

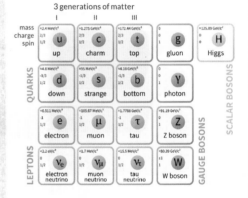

Standard model of elementary particles

The Standard Model explains almost everything about the way the Universe behaves at a fundamental level.

2/ Shortcut: The Standard Model is a successful theory because many of its predictions have been confirmed. For instance, in the 1960s six physicists including **Peter Higgs** predicted the existence of a force-like field permeating the Universe alongside the fields formed by the four known forces. This "Higgs field" could explain why some particles have mass. The existence of the Higgs field could be confirmed by finding its associated boson: the famous Higgs boson. In 2012, almost fifty years after the Higgs boson was proposed, physicists at the Large Hadron Collider discovered a particle with many of its predicted properties.

See also //

66 The special theory of relativity, p.136

Peter Higgs // b. 1929

3/ Hack: The Standard Model is a successful and relatively simple theory that does a great deal.

It gives a fundamental explanation for matter, mass, and three of the four forces that govern the observable Universe.

No.88
String theory

A true theory of everything?

1/ Helicopter view: The **Standard Model** (see page 178) uses quantum physics to provide an almost complete explanation for the way the observable Universe behaves at a fundamental level. But it falls down when it comes to one of the most familiar of all physical forces: gravity.

Gravity was the first of the fundamental forces operating in the Universe to be studied in any great detail. At the subatomic scale, though, gravity is weak, and its mode of operation is still a mystery. In principle it might behave rather like the other fundamental forces, which means it should be carried between objects via a special kind of subatomic particle dubbed the graviton. But physicists have yet to find strong evidence that gravitons exist. Many physicists think a new theory might be needed to explain gravity in terms of quantum physics. **String theory** is one of the leading candidates.

String theory emerged gradually in the second half of the 20th century. It imagines that the elementary particles – things like electrons and quarks (see: **The quark model**, page 176) – are actually one-dimensional vibrating filaments, giving them the appearance of tiny lengths of string.

It is a complicated theoretical framework that is mathematically consistent only if physicists assume that there are many extra spatial dimensions beyond the three we are familiar with. For this reason, string theory is conceptually difficult to understand. But it does seem to offer a quantum explanation for gravity: one of the ways strings are theorized to vibrate gives them many of the expected properties of the elusive graviton. Even so, there is no consensus yet on whether some version of string theory really is a theory of everything. Physicists still have much to discover.

Physicists resort to exotic algebraic geometry to study string theory's many dimensions.

2/Shortcut: String theory has evolved a great deal over the years, gaining in complexity and multiplying so that, by the 1990s, there were several distinct versions. In 1995, **Edward Witten** found a way to show that all of these distinct versions were actually part of one all-encompassing theory dubbed "M-theory". If M-theory is correct, all of the known elementary particles must have what is called a "superpartner" – which essentially suggests that there is another, yet to be discovered set of elementary particles. But as of mid-2017 there was no sign of their existence – possibly a sign that M-theory needs a rethink.

See also //

56 Newton's law of universal gravitation, p.116

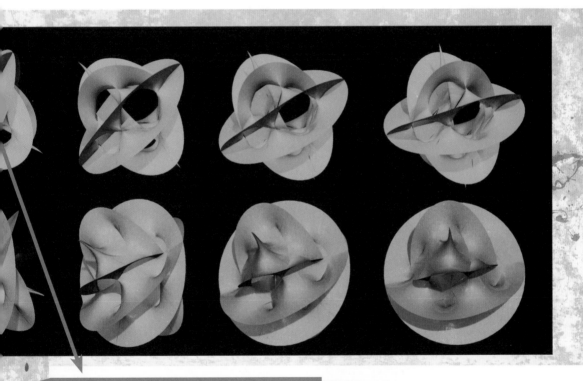

3/Hack: String theory is intellectually challenging to understand, with its insistence that the Universe contains ten spatial dimensions rather than the three we are familiar with.

But its complexity might be worth it: it might offer a full quantum explanation for the way the Universe works.

No.89
The concept of metallicity
What is inside the Sun?

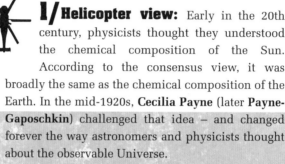

Cecilia Payne-Gaposchkin //
1900–1979

I/ Helicopter view: Early in the 20th century, physicists thought they understood the chemical composition of the Sun. According to the consensus view, it was broadly the same as the chemical composition of the Earth. In the mid-1920s, **Cecilia Payne** (later **Payne-Gaposchkin**) challenged that idea – and changed forever the way astronomers and physicists thought about the observable Universe.

In the 19th century, **William Hyde Wollaston** and **Joseph von Fraunhofer** had noticed that the light from the Sun, when broken up into its rainbow-like spectrum, was interrupted by a series of thin black bars. It was as if specific colours (wavelengths) of light coming from the Sun had gone missing on the way to Earth, leaving bars of darkness in the spectrum.

Later in the century, **Gustav Kirchhoff** and **Robert Bunsen** came up with an explanation. They heated various substances in the lab and established that specific chemical elements emit (and absorb) light of very precise wavelengths. Physicists began to accept that the Sun's "missing" light could be explained if some of these chemical elements were present in the Sun's atmosphere, absorbing specific wavelengths of sunlight. Metals like calcium and iron seemed particularly abundant in the Sun. This was seen as significant given that these metals are also relatively abundant in the Earth's crust.

But Payne argued that the physicists were reading the solar data incorrectly. Early in the 1920s, she had met **Meghnad Saha**, whose research strongly suggested that the temperature of a chemical element influences the way it absorbs light. Factoring in that the solar atmosphere is extremely hot, Payne argued that the Sun is composed mostly of hydrogen and helium.

Payne's work led to the **concept of metallicity** – the idea that *all* stars are mainly hydrogen and helium, and contain only small amounts of other elements (all of which are labelled "metals"). It was an important clue to understanding the way the Universe formed.

The Harvard computers helped make sense of the observable Universe.

2/Shortcut: Late in the 19th century, **Edward Charles Pickering** began hiring women to help analyse astronomical data. The women became known as the "Harvard Computers". One – **Annie Jump Cannon** – is recognized for the intricate star classification scheme she devised. Cannon grouped stars based on the features of their light spectra. At face value, Cannon's scheme seemed to suggest that stars were highly variable in their chemistry. Payne argued that they were not: her research suggested all stars contained mostly hydrogen and helium, and that Cannon's categories highlighted differences in star *temperature* rather than chemistry. With time, Payne's idea became widely accepted.

See also //

92 The big bang theory, p.188

3/Hack: Stars come in a wide variety of forms: red supergiants can be a thousand times larger than the Sun; Wolf-Raylet stars can be 30 times hotter.

But, compositionally, all stars are dominated by hydrogen and helium.

No.90
The cosmic
distance ladder A stairway to
the stars

1/ Helicopter view: In the early 1920s, the scientific consensus was that the Universe was large – but not *that* large. Leading astronomers including **Harlow Shapley** assumed that there was little, if anything, beyond our Milky Way Galaxy. **Edwin Hubble** helped convince the research community that Shapley was wrong.

Hubble's work leaned heavily on an observation made a few decades earlier by **Henrietta Swan Leavitt**. She had studied Cepheid variables – unusual stars that brighten and dim over time – in two regions of the night sky called the Magellanic Clouds. Leavitt assumed that the Cepheid variables in each Magellanic Cloud were roughly the same distance from Earth, and she noticed that the brighter the star, the slower its "period" – the rate that it brightened and dimmed.

In the 1910s, **Ejnar Hertzsprung** realized Leavitt's discovery was hugely significant. Until that time it was possible to measure the distance between Earth and stars that are relatively near to the Solar System using a technique called stellar parallax. Hertzsprung argued that Cepheid variables could serve as an additional tool for measuring greater astronomical distances. With time, even more measurement techniques were added, leading to what astronomers now call the **cosmic distance ladder** – a system that helped estimate the size of the Universe.

Back in 1924, Hubble made an important contribution to the ladder's development. He used Hertzsprung's work to estimate the distance of Cepheid variables in two nebulae in the sky. The figures he came up with – about 900,000 light years (8.5 quintillion kilometres/5.9 quintillion miles) – were vast. He concluded that the two nebulae must lie far beyond our Milky Way. Both are now recognized as distinct galaxies. Shapley initially dismissed Hubble's work as "junk science". But soon both he, and other astronomers, came to accept Hubble's calculation, and its implication that the Universe was astonishingly large.

Leavitt's work led to the cosmic distance ladder that established the size of the observable Universe.

2/ Shortcut: Some stars seem to shift in the sky as our planet orbits the Sun. This movement ("stellar parallax") is used to calculate their distance – but this method is only useful for stars relatively near Earth. Cepheid variables help establish greater cosmic distances. The *measured* brightness of a star weakens the further it is from Earth. But a Cepheid's "period" relates to its *actual* brightness. By using stellar parallax to measure the distance to nearby Cepheids, astronomers established a simple relationship between a Cepheid's actual and measured brightness, and its distance to Earth. They could then work out the distance to remote Cepheids.

See also //

91 Hubble's law, p.186

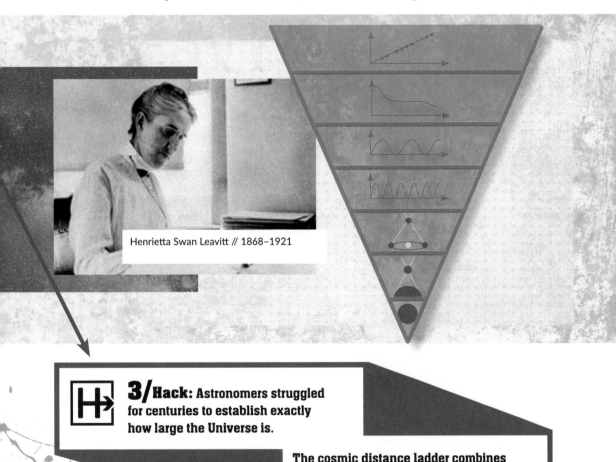

Henrietta Swan Leavitt // 1868–1921

3/ Hack: Astronomers struggled for centuries to establish exactly how large the Universe is.

The cosmic distance ladder combines several complementary techniques to measure further and further into space.

No.91
Hubble's law

The Universe becomes dynamic

Edwin Hubble //
1889–1953

1/Helicopter view: Scientists of the early 20th century could not agree on the size of the Universe, but on one point there was a broad consensus. The Universe was static. Then, at the end of the 1920s, **Edwin Hubble** published findings that led to a scientific change of mind.

The assumption of a static Universe underpinned the work of even the most famous scientists. When **Albert Einstein** tried to model the (static) Universe using his **general theory of relativity**, he found that he had to insert a new term into his equations – dubbed the "cosmological constant" – to prevent the Universe collapsing in on itself due to gravity.

Some physicists were dissatisfied by Einstein's solution. In 1922, for instance, **Alexander Friedmann** suggested Einstein should consider the possibility of a dynamic Universe. **Georges Lemaître** made a similar argument in 1927. Einstein rejected the idea.

Within a few years, though, Einstein changed his mind. In 1929, Hubble – fresh from his discovery of stars beyond the Milky Way (see: **The cosmic distance ladder**, page 184) – made an even more astonishing observation. He argued that these distant stars were moving away from Earth, and the greater the distance they were from our planet, the *faster they were receding*. Other scientists soon accepted this relationship. It became known as **Hubble's law**.

The observation convinced many, including Einstein, that the Universe must be expanding. Within a few years, he had abandoned the idea of the cosmological constant. Famously, Einstein is claimed to have later declared the cosmological constant his "biggest blunder". However, some science historians have cast doubt on the veracity of this quote. What's more, some recent discoveries have led to a renewed interest in the cosmological constant (see: **The accelerating Universe theory**, page 192).

The further a star is from Earth the faster it is speeding from us – implying an expanding Universe.

2/Shortcut: Hubble analysed the light spectra arriving on Earth from distant stars. There were characteristic lines in these spectra, due to the presence of chemical elements – particularly hydrogen – in each star's atmosphere. However, these spectral lines did not appear at exactly the same frequencies as they do in the light reaching Earth from our Sun. Instead, the spectral lines in the starlight always lay further toward the red end of the light spectrum than they *should* have done. The **special theory of relativity** predicts this "redshift" should occur to light leaving an object that is traveling away from an observer, as a consequence of the distortion of spacetime. Hubble discovered that the further away a star appeared to be, the greater the redshift (implying the faster it was speeding away from Earth). This suggested the Universe was expanding.

See also //

66 Special theory of relativity, p.136

68 General theory of relativity, p.140

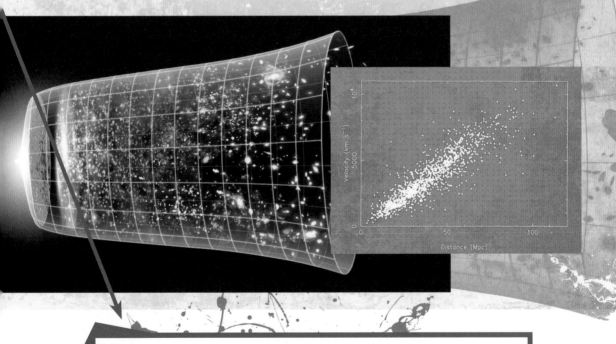

3/Hack: Physicists in the early 20th century, including Einstein, assumed that the Universe was static.

Hubble's careful observations helped convince most of them that the Universe is, in fact, expanding.

No.92
The big bang theory
The Universe gains a beginning

1/Helicopter view: Physicists the world over were astonished when, in 1929, **Edwin Hubble** published evidence that strongly suggested the Universe is expanding. One physicist – **Georges Lemaître** – must have been especially excited by the development. He had predicted an expanding Universe a few years earlier. In 1931, Lemaître built on his idea, and inadvertently triggered a decades-long scientific controversy.

If the Universe is expanding, Lemaître reasoned that it must have been *smaller* in the past. Logically, he argued, there was a point long, long ago when all the mass in the Universe must have existed in the form of a single atom.

This idea was deeply unpopular. It seemed uncomfortably like an appeal to a God-like creation event. The fact that Lemaître was a Roman Catholic priest probably did not help. Instead, many physicists preferred what became known as the **steady state theory**, which essentially suggested that the Universe had no beginning and will have no end. Although Hubble's work had convinced many scientists that the Universe is expanding, steady state theorists argued that, if matter were constantly being generated as this expansion continued, the Universe itself would never really change its appearance.

Not all were convinced, however. In the late 1940s, **Ralph Alpher** and **Robert Herman** made a series of predictions based on the idea that the Universe had been very small and very hot in its earliest stages. The two researchers suggested that the "afterglow" from this early stage should still linger in the Universe as a subtle radiation signal.

In 1964, **Arno Penzias** and **Robert Wilson** found just such a radiation signature – now dubbed the "cosmic microwave background". Its discovery was the final nail in the steady state theory. **Fred Hoyle** had given Lemaître's alternative theory a striking name – the "big bang" – in 1949. By the late 1960s the **big bang theory** was becoming the consensus view.

The work of Lemaître and Hubble helped lead to acceptance of the big bang theory.

2/ Shortcut: According to the steady state theory, the Universe has always looked as it does now. But in 1961, two astronomers – **Martin Ryle** and **Randolph Clarke** – realized that very distant regions of the Universe contained an unexpectedly large number of unusual objects called "radio sources". These distant regions of space are older than nearer regions (the light from them has had to travel much further to reach Earth, so it has taken longer to do so). As such, Ryle and Clarke's discovery had an important implication. In the distant past, the Universe contained more radio sources than it does today: it once *looked different*. The discovery was a blow for the steady state theory, and paved the way for the rise of the big bang theory.

See also //

91 Hubble's law, p.186

Georges Lemaître // 1894–1966

3/ Hack: The big bang theory is such a bold idea that many scientists were initially sceptical. But as they continued to study space through their telescopes, it began to dawn on them:

The appearance of the observable Universe makes sense only by assuming it began with a big bang.

No.93
The concept of dark matter An 85-year-old (and counting) mystery

1/Helicopter view: As early as the 19th century some physicists suspected that there might be literally more to the Universe than meets the eye. For many scientists, though, the **concept of dark matter** really began with **Fritz Zwicky** in 1933.

Zwicky was studying the Coma Cluster – a collection of galaxies so densely packed in a small region of space that they are loosely bound together by gravity. Zwicky measured the speed of eight of the galaxies in the cluster. The figures he obtained – roughly 1,000 kilometres (600 miles) per second – were a surprise: at that speed, the galaxies should have been able to escape the gravitational pull of the cluster and fly off.

Either there was something wrong with **Newton's law of universal gravitation** or the Coma Cluster contained a lot more mass than its visible matter suggested – enough to generate sufficient gravity to pull the fast-moving galaxies together. Zwicky called the missing mass "dunkle (kalte) Materie" – dark (cold) matter.

In 1936, **Sinclair Smith** noticed that a different galactic cluster – the Virgo Cluster – also seemed to have a higher mass than suggested by its visible matter. And 1939 brought yet more evidence, this time from **Horace Babcock**'s study of the Andromeda Galaxy.

Puzzlingly, though, these observations were more or less ignored until the early 1970s. There then followed a relatively intense re-analysis of the problem using new measurements – and the results essentially confirmed the observations made in the 1930s. By the late 1970s there was general consensus that the observable Universe does, indeed, contain a lot more matter than astronomers can detect – but there was no agreement on exactly what *form* the missing matter took.

The mystery remains unsolved. Most scientists accept that dark matter comprises roughly 85 per cent of the mass of the observable Universe – but their best efforts have so far failed to identify what it is.

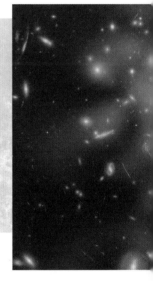

Galaxy clusters may be held together by dark matter.

2/Shortcut: Computer simulations built by **Jeremiah Ostriker** and **James Peebles** in the 1970s suggest our Milky Way Galaxy can only maintain its shape if it has a massive "halo" of dark matter extending well beyond its visible edge. Assuming that this dark matter halo comprises unusual subatomic particles that interact to a very limited degree with typical matter particles, countless such particles should pass through Earth every second. Several laboratories have been built to detect the subtle effects these hypothetical "weakly interacting massive particles" (WIMPs) should have on typical atoms as they travel through the planet. But to date, none of these experiments have produced convincing evidence that WIMPs really exist.

See also //

56 Newton's law of universal gravitation, p.116

Fritz Zwicky // 1898–1974

3/Hack: Many galaxies, and galaxy clusters, seem to have a much stronger gravitational field than the visible matter they contain can generate. The leading explanation for the discrepancy is that they contain huge amounts of invisible dark matter.

But no one knows for sure what form this hypothetical dark matter takes.

No.94
The accelerating Universe theory
Why a "big crunch" now seems unlikely

 1/Helicopter view: Just ninety years ago, many scientists assumed that the Universe never really changes its appearance. Late in the 1920s, **Edwin Hubble** helped convince many of them that they were wrong and that the Universe is actually expanding. Right at the end of the 20th century, some astrophysicists added a twist to the idea. They argued that the observable Universe is expanding at an *ever-faster rate*.

The **big bang theory** argues that the observable Universe has been expanding for billions of years from an exceptionally tiny initial volume. But most scientists assumed that the expansion rate should eventually slow down. All of the matter spread through space is mutually attracted (see: **Newton's law of universal gravitation**, page 116). Eventually, the argument went, the gravitational pull acting across all matter should decelerate the rate of the Universe's expansion. It was even conceivable that the observable Universe contains enough matter to eventually reverse the expansion, beginning a contraction of space that would end in a "big crunch".

An international scientific team including **Brian Schmidt** and **Adam Riess** began looking for evidence of this deceleration. They focused on extraordinarily bright supernovae in the deeper parts of the observable Universe. They knew how bright the supernovae *should* appear, assuming a decelerating Universe – but the supernovae actually turned out to be much dimmer (see: **The cosmic distance ladder**, page 184). The obvious explanation, they argued, was that the supernovae were further away than the astrophysicists had assumed them to be. The observable Universe must be expanding at an accelerating rate to have flung them so far from Earth.

The research community quickly accepted the observations as key evidence for an **accelerating Universe theory**. They realized it implied the existence of some mysterious Universal "antigravity" that drives the acceleration. This is sometimes called "dark energy".

Brian Schmidt // b. 1967

Galaxies appear to be moving away from each other at an ever faster rate.

2/Shortcut: Not only do most physicists accept the accelerating Universe theory, they think that **Albert Einstein** inadvertently predicted it a century ago. In 1917, Einstein had discovered that the equations of his **general theory of relativity** still worked if he inserted an additional term he called the "cosmological constant" (see: **Hubble's law**, page 186). Although Einstein abandoned the term in the 1930s, some of today's physicists think he should not have done. If the cosmological constant is given a positive value, it arguably *predicts* an accelerating expansion of the Universe. Some physicists think Einstein's cosmological constant can be viewed as the earliest indication of the existence of "dark energy".

See also //

68 General theory of relativity, p.140

91 Hubble's law, p.186

92 Big bang theory, p.188

3/Hack: At the end of the 20th century many physicists assumed that the big bang-driven expansion of the observable Universe would eventually slow down or even reverse.

The accelerating Universe theory suggests the expansion may continue to pick up speed forever.

No.95
The inflation theory
A surprise at the birth of the Universe

 1/ Helicopter view: Several bold new theories about the Universe emerged in the early years of the 20th century. Many of the predictions these theories made were confirmed through observation – but not all of them. It was as if the theories were missing a vital piece. Some physicists argue that **inflation theory** provides that missing piece.

Most physicists think that, in its earliest history, the Universe was incredibly hot, dense and small (see: **The big bang theory**, page 188). Explaining exactly how this young, hot Universe behaved is a significant 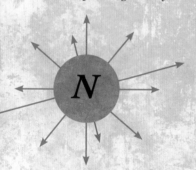 challenge, but physicists have been developing ideas that they think might hold important clues (see: **String theory**, page 180). There is a significant problem with these theories, though. Many of them predict that the early Universe should have given rise to an abundance of strange and exotic subatomic particles called magnetic monopoles, and that these particles should remain abundant in the Universe to this day. However, the observable Universe appears to be monopole-free.

In the late 1970s and early 1980s, **Alan Guth** developed an idea that could explain the missing monopoles. He argued that, within the first second of the Universe's existence, it expanded at an astonishing rate – far faster even than the speed of light. It then snapped out of this brief "inflationary epoch" and began behaving as it does today. If this inflation occurred immediately *after* the Universe became flooded with monopoles, the process should have drastically diluted their density. Crudely speaking, it would be akin to the way that currants in a sweet bread dough move further from each other as the dough rises.

Guth later realized that his idea also explained other profound mysteries, including one called the horizon problem. Because of these successes, inflation theory has become popular – although it has several high profile and vocal critics.

The observable Universe has a magnetic monopole problem – inflation theory offers a solution.

2/Shortcut: Imagine dropping a water heater into a cold swimming pool and then discovering that all of the water in the pool is warm just ten seconds later. It sounds impossible: the water at the far end of the pool needs more time to respond to the heater. The horizon problem is a little like this: the properties on one side of the observable Universe are similar to those on the opposite side, even though they are too far apart to have responded to each other since the Universe began. Inflation theory offers a solution. Imagine dropping the heater into a glass of water, which – after a few seconds – inflates to the size of a swimming pool. In those few seconds before inflation, the water gained a uniform temperature – and then maintained that temperature as it expanded.

See also //

97 The Multiverse hypotheses, p.198

Alan Guth // b. 1947

3/Hack: Late in the 20th century, physicists realized that some of their observations were not predicted by their theories of the Universe.

They found they could solve the problems if the Universe had briefly expanded in size at an extraordinary rate early in its history.

No.96
The Goldilocks Universe concept
A reality made for humans?

1/ Helicopter view: We are lucky. The Earth orbits the Sun at a distance that allows for liquid water on its surface – widely seen as a prerequisite for life as we know it. But many physicists think our fortune is much more profound than this. They point out that the observable Universe is curiously well suited for the existence of matter, stars, planets and life as we know it. Our Universe, they say is *just right*: it is a Goldilocks Universe.

Many physicists over the last fifty years have commented on the fact that the fundamental properties of the observable Universe are seemingly fine-tuned for life. In the late 1980s, **Stephen Hawking** noted that a slightly different value on the electric charge of the electron (see: **The plum pudding model**, page 148) would prevent stars functioning in a way that ultimately provided the chemical elements used by life on Earth.

Martin Rees also explored the idea in a 2001 book. He highlighted half a dozen physical constants – basically, six numbers – that govern how everything from atoms to galaxies behave. As far as physicists can tell, these six constants are entirely unconnected to one another. But all six are, independently, ideally suited for life as we know it. A small change to any one of them could have led to a Universe that collapsed long ago, or prevented the formation of galaxies. Again, the implication is that the Universe is "fine-tuned" for life.

Even so, the **Goldilocks Universe concept** sits uncomfortably with many physicists, who worry that it could be seen to hint at the guiding hand of a creator. Some scientific theories offer a rational explanation for our finely-tuned Universe (see: **The multiverse hypotheses**, page 198).

Not too hot, not too cold, the observable Universe is just right for life as we know it.

2/ Shortcut: The Goldilocks Universe concept is arguably as much about philosophy as science. The observable Universe in its current form is suitable for carbon-based life – life as we know it. But some scientists argue that life could exist in forms we cannot imagine. A Universe with slightly different properties might not be able to provide the conditions for carbon-based life, but it might still support intelligent living things. In a sense, those alternative Universes would also be fine-tuned to life – just not carbon-based life. To put it another way, perhaps it is not surprising that life on Earth is carbon-based given that the preconditions of the observable Universe allow for that possibility. Maybe it is not our Universe that is fine-tuned for life, but carbon-based life that is fine-tuned to the Universe.

See also //

92 The big bang theory, p.188

Martin Rees // b. 1942

3/ Hack: Physicists have worked out a series of complex equations that explain the behaviour of the Universe.

Tweaking some of the numbers would have sent our Universe down a very different path unsuitable for carbon-based life.

No.97
The Multiverse hypotheses

Is there anything else out there?

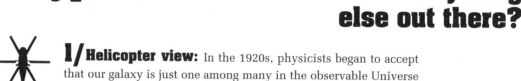

1/ Helicopter view: In the 1920s, physicists began to accept that our galaxy is just one among many in the observable Universe (see: **The cosmic distance ladder**, page 184). Will we one day discover that the observable Universe itself is just one of a countless number of other universes in existence?

The observable Universe is – predictably – the region of outer space we can observe. As such, it is finite in size, because the light we observe is thought to travel through space at a finite speed (see: **The special theory of relativity**, page 136), and the Universe is thought to have a finite age (see: **The big bang theory**, page 188). In other words, there is an accepted limit on the distance light can have traveled to reach Earth since the first stars formed.

But there might be vast regions of space beyond the "cosmological horizon" of our observable Universe. Interest in this idea picked up in the 1980s with the suggestion that the Universe underwent a brief interval of extraordinarily fast expansion immediately after its formation (see: **The inflation theory**, page 194). Physicists including **Paul Steinhardt** and **Andrei Linde** realized that, theoretically, the process of inflation never ended. As such it is theoretically possible that the region of space beyond the cosmological horizon is *inconceivably large* and contains a multitude of other "universes" – collectively making a "Multiverse".

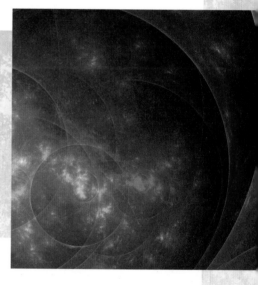

There are several other **Multiverse hypotheses**. All are viewed as problematic by some physicists because observational evidence of these hypothetical other universes may be impossible to obtain. Other physicists are unconcerned by this point. They argue that if a theory explains features of the observable Universe and also implies the existence of other universes, we should accept those universes exist even though we may never see them.

Are there other universes out there waiting to be discovered?

2/Shortcut: One Multiverse theory suggests there might be an inconceivable number of universes in the vastness of space. Some may look identical to ours – but more importantly, some might look very different, ruled by different versions of the laws of physics. This might explain a mystery: why our Universe seems to be governed by laws curiously tuned for the existence of carbon-based life. If there is a vast number of universes in existence it might be simple probability that one – ours – has a version of the laws that allow for human life. However, if this idea is correct it implies that there is no point trying to establish *why* our Universe has the physical laws it does. It might be a statistical accident. Many physicists (including Steinhardt) don't like this idea.

See also //

98 The many-worlds interpretation, p.200

Andrei Linde // b. 1948

3/Hack: Our observable Universe is vast – 93 billion light years across – but it might be a tiny speck in an inconceivably larger reality.

This reality might contain innumerable other regions broadly analogous to our observable Universe. Collectively, these "universes" would form a Multiverse.

No.98

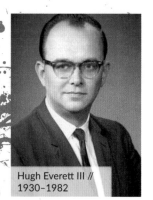

Hugh Everett III //
1930–1982

The many-worlds interpretation
Are there alternative versions of reality?

1/Helicopter view: Physicists have now begun to consider the possibility that our observable Universe is just one of many (see: **The Multiverse hypotheses**, page 198). Arguably, however, these hypothetical additional universes are relatively mundane – they could be viewed as a collection of islands in a single sea of reality. The realm of quantum physics offers something much more exotic.

In the 1950s, many physicists were deeply troubled by the predicted outcomes of quantum physics. **Hugh Everett III** was particularly concerned by what has become known as the "measurement problem", made famous by **Erwin Schrödinger** in a 1930s thought experiment.

Put simply, behaviour at the subatomic – quantum – scale seems to be governed by probability. A particle can simultaneously exist in one of several distinct forms at different levels of probability (for instance, fifty per cent chance it is in form A, fifty per cent chance it is in form B) until the moment it is observed in some way. At that point the probability "wave function" collapses and the particle takes on one unambiguous form: one hundred per cent chance it is in form A. What, wondered Everett, happens to those other possible versions of the particle at the moment the wave function collapses?

He came up with an extraordinary idea. Perhaps the wave function *never* collapses. Perhaps it continues to branch away into different versions of reality, none of which can communicate with any of the others. In each of these realities there is a distinct version of the experimental observer (and, by extension, a distinct version of Earth, our Milky Way Galaxy, and so on).

The idea was largely ridiculed – and then ignored for a few decades. But in the 1960s and 70s, **Bryce DeWitt** (originally sceptical of Everett's idea) revived and popularized it. DeWitt named the idea the **many-worlds interpretation**. It retains champions – and critics – to this day.

The many-worlds interpretation suggests new doors to different realities are always opening.

2/ Shortcut: Schrödinger's cat is, famously, both alive and dead. More specifically, it has some probability of being alive and some probability of being dead – until an observer peers inside the box containing the cat and collapses the probability wave function. But Everett argued that even then the wave function does not collapse. An observer in our world may confirm the cat is alive, but at that precise moment another parallel version of reality pops into existence in which the observer finds a dead cat inside the box. The catch is that we cannot communicate with these parallel worlds and so, argue "many-world" critics, we cannot confirm they exist.

See also //
79 Schrödinger's cat, p.162

3/ Hack: Subatomic objects seem to have the potential to exist in several states simultaneously, until they are observed.

The many-worlds interpretation argues that they continue to exist in many states even then.

No.99
The big splash hypothesis   The birth of the Moon

1/ Helicopter view: The Sun has long been the focus of intense scientific interest, while the Moon has received less attention. However, in the last forty years, scientists have come to a broad consensus on how Earth's natural satellite formed. It began, they argue, with a big splash.

George Darwin //
1845–1912

Interest in the Moon's formation stretches back at least as far as **George Darwin** (**Charles Darwin**'s fifth child). In the 1870s, Darwin suggested that our planet spun very rapidly when it was young – so rapidly that material was flung off, eventually coalescing into the Moon. The idea, which became known as the **fission hypothesis**, fell out of favour early in the 20th century. **Forest Ray Moulton** and others pointed out that a Moon formed this way would lack energy – it would probably fall back to Earth in due course. Although the fission hypothesis was revived in the 1960s, a new model was already emerging that would prove more popular.

In 1946, **Reginald Daly** suggested that early Earth could have collided with a planetoid, and that material from the impact might then have coalesced into the Moon. **William Hartmann** and **Donald Davis** developed a similar idea in the mid-1970s after considering how the planets in the Solar System formed in the first place. They argued that relatively large planets like the Earth were originally accompanied by a series of smaller "planetesimals" in roughly the same orbit. Hartmann and Davis calculated that a planetesimal about 1,200 kilometres (746 miles) in diameter would, if it hit the Earth, impact with enough energy to kick material far from the planet – material that could ultimately coalesce to form the Moon.

The idea struck most scientists as plausible, and it became the favoured hypothesis for the Moon's formation. It is now known as the **giant-impact hypothesis** – or, more informally, as the **big splash hypothesis**.

Many scientists think the Moon appeared after a huge cosmic collision.

2/Shortcut: Most scientists now accept the idea that the Earth collided with a hypothetical planetesimal – sometimes dubbed Theia – when it was just a few tens of millions of years old. They argue that relatively low-density rocky material from Earth (and Theia) was ejected into space while most of the denser, iron-rich material sank into the Earth's core (see: **The dynamo theory**, page 80). Within a few decades, a lot of that ejected rocky material coalesced to form the Moon. This big splash hypothesis explains many of the Moon's features, including its relatively low density, and the fact that lunar rocks collected during the Apollo missions have an Earth-like chemical signature.

See also //

89 The concept of metallicity, p.182

3/Hack: Scientists argue that the Earth formed about 4.54 billion years ago, and gained its Moon just a few tens of millions of years later.

The leading hypothesis is that the Moon formed when the Earth collided with another planet-like body in a "big splash".

No.100
The panspermia hypothesis
Where did we really come from?

1/ Helicopter view: The scientific advances of the last four hundred years have revolutionized our understanding of the Earth and the Universe. But there are questions that remain unanswered, including one of the most fundamental questions of all: where did we come from?

Most biologists think humans originally came from Africa a few million years ago (see: **The out of Africa hypothesis,** page 68). Most also argue that all living things can trace their roots back billions of years (see: **The LUCA hypothesis,** page 32). But what about before then?

The **biogenesis hypothesis** argues that life (even single-celled organisms like bacteria) comes from life. Some 19th century researchers, including **William Thomson** (later **Lord Kelvin**) argued on inductive grounds (see: **The black swan problem**, page 72) that biogenesis is a universal law, like **Newton's law of universal gravitation**. If life always comes from life, they argued, then life must *always* have existed: it had no beginning.

Svante Arrhenius //
1859–1927

However, Thomson suspected that the Earth itself was relatively young (see: **The concept of radiometric dating**, page 170), which suggested to him that life on Earth must have arrived from outer space on meteorites. The idea became known as the **panspermia hypothesis**.

Even in the 19th century the idea was controversial. It remains so to this day. Significantly, the present day consensus view is that the Universe *did* have an origin, roughly 13.8 billion years ago (see: **The big bang theory**, page 188). As such, a central assumption of Thomson's version of the panspermia hypothesis is now seriously questioned: if the Universe is not eternal, neither is life. It must have sprung from inorganic matter at least once in the Universe. Many biologists suspect life arose from "non-life" on early Earth – and they are making progress toward understanding how this might have happened (see: **The RNA world hypothesis**, page 50).

Some scientists think life on Earth actually originated elsewhere in the Universe.

2/Shortcut: There are several versions of the panspermia hypothesis, but all of them assume that at least some organisms can travel through the hostile environment of outer space. **Svante Arrhenius** was one of the first scientists to really explore this idea. Early in the 20th century he suggested on theoretical grounds that organisms the size of bacteria, if they escaped Earth's gravitational pull, would be "pushed" by radiation pressure from the Sun with such force that they could reach a nearby star system – Alpha Centauri – in about nine thousand years. Arrhenius also cited experiments suggesting that bacteria can survive not only the extreme radiation they would be exposed to in space, but also the extraordinarily low temperatures they would experience there.

See also //

47 The biogenesis hypothesis, p.98

56 Newton's law of universal gravitation, p.116

3/Hack: Scientists have not yet established with confidence how life on Earth came into being.

The panspermia hypothesis suggests we don't even know where in the Universe life on Earth originated.

Index

Acknowledgements

Author's acknowledgements

In writing this book I pulled together information from many sources: peer-reviewed scientific papers published in the academic literature as well as online resources including the Stanford Encyclopedia of Philosophy, UC Berkeley's Understanding Evolution website, the 'This Month in History' column on the APS News website, the websites of *New Scientist*, *Scientific American* and the *Guardian*, BBC Earth and BBC Radio 4's *In Our Time* series.

I would like to acknowledge in particular the following books: *Convergent Evolution: Limited Forms Most Beautiful*, G. R. McGhee; *Growing Young* (2nd edition), A. Montagu; *Earth Science: The People Behind the Science*, K. E. Cullen; *The Panda's Thumb*, S. J. Gould; *Time's Arrow, Time's Cycle: Myth and Metaphor in the Discovery of Geological Time*, S. J. Gould; *James Lovelock: in Search of Gaia*, J. Gribbin and M. Gribbin; *Discovering the Ice Ages*, T. Krüger; *Georges Cuvier, Fossil Bones, and Geological Catastrophes*, M. J. S. Rudwick; *North Pole, South Pole: the Epic Quest to Solve the Great Mystery of Earth's Magnetism*, G. Turner) *Inventing Temperature: Measurement and Scientific Progress*, H. Chang; *Modern Physics for Scientists and Engineers* (2nd edition), S. T. Thornton and A. Rex; *Quantum Reality – Theory and Philosophy*, J. Allday (; *Physics and Beyond – Encounters and Conversations*, W. Heisenberg; *Just Six Numbers: the Deep Forces that Shape the Universe*, M. Rees; *The Nature of Consciousness, the Structure of Reality*, J. D. Wheatley.

Picture/artwork credits

Alamy Stock Photo 19th era 41r; AA Images 110; Callista Images/Cultura Creative (RF) 49c; Chronicle 100; Classic Image 14; Corbin17 35; David Keith Jones/Images of Africa Photobank 58; Gary Doak 179; GL Archive 99l; Granger Historical Picture Archive 85; Grant Heilman Photography 145r; Ivan Vdovin 41al; Kevin Schafer 96; Keystone Pictures USA/ZumaPress.com 176; Laguna Design/Science Photo Library 180; Liam Bailey 197; Liam White 46r; LOC 91l; Mediscan 104; National Geographic Creative 195; Neil Spence 92; Nobumichi Tamura/Stocktrek Images 34r; Phil Degginger 105c; Photo Researchers/Science History Images 27c, 42, 55c, 132; Photo12/Ann Ronan Picture Library 169bc; Pictorial Press Ltd 82l, 104, 148, 167; Science History Images 127l, 172, 186, 187, 189cr; Sputnik 199; Steve Gschmeissner/Science Photo Library 113r; The Print Collector 44; The Protected Art Archive 84; Vintage Power and Transport/Mark Sykes 127r; World History Archive 43r. **Courtesy of Mark Everett** (Source: http://ucispace.lib.uci.edu/handle/10575/1060) 200. **Courtesy United Nations** (Ray Witlin) 60. **Dreamstime.com** Carmen Craig 19bg; Maxirf 11l, Neosiam 116; Pavel Konovalov 162; Thomas Jurkowski 166. **Getty Images** 23, © Ted Streshinsky/CORBIS/Corbis via Getty Images 65r; Bettmann 191; Onas Ekstromer/AFP 192; Yoon S. Byun/The Boston Globe via Getty Images 62. **Gitschier J** (2010) All About Mitochondrial Eve: An Interview with Rebecca Cann. PLoS Genet6(5): e1000959 (https://doi.org/10.1371/journal.pgen.1000959) 48. **Library of Congress, Prints and Photographs Division** 134, 142, 168. **NASA** 205r; JPL-Caltech 203; ESA and D. Coe (NASA JPL/Caltech and STScI) 189r, 190; The Hubble Heritage Team and A Riess (STSci) 193. **NOAA** 33 left. **Reproduced by permission of the Rockefeller Archive Center** 28. **Shutterstock** adike 107; Africa Studio 111r; Alemon cz 97c; Alexander Raths 101c; Alexey Godzenko 101bg; alexialex 108; alexokokok 89c; Alila Medical Media 39l; Andrii Muzyka 41cl; Andrii Vodolazhskyi 106bg; Andromed 47; Andrzej Kubik 59l; aniad 4; Asit Jain 65l; attaphong 133; AVA Bitter 37l; Belish 184; Bernhard Staehli 89bg; betibup33 177bg, 198; Bjoern Wylezich 171bl; Carlo2017 98r; catshila 103bg; Chatree.l 103c; cigdem 81; concept w 169l, 169r, 170r; Cornel Constantin 19c; Cristina Rabascall 99r; Distinctive Shots 121; DmitriyRazinkov 39c; Dominionart 139r; dondesigns 95; DR Travel Photo and Video 94bg; Dutourdumonde Photography 63ac; Eladora 67; Eric Isselee 17; Everett Historical 78, 86, 98l; Frederick J. Horne 96bg; Galyamin Sergej 82r, 153; general-fmv 139l; Georgios Kollidas 77 inset; gerasimov_foto_174 59r; Green Flame 15l; Hayati Kayhan 137l; Hein Nouwens 13; Igor Zh. 77; iryna1 117, 118; Jaimie Tuchman 9r; Jan Kaliciak 51c;, janez volmajer 111l; Janis Abolins 53bl, 53br; jannoon028 131l; Jenn Huls 201; Juan Gaertner 109bg; Khamkhlai Thanet 29bg, 29c; Klaus Balzano 91bg; ldambies 119; LongQuattro 12ac, 19b, 41br; Lynn Y 49bg; magnetix 149; mamita 11r; Mark Godden 74; mark higgins 73; MikhailSh 25l; mkarco 115r; molekuul_be 114bg; Morphart Creation 12ar, 27 inset, 129l, 129r, 131r; nicemonkey 5b, 5r, 7b, 18l; Nostalgia for Infinity 171al; nuu_jeed 56; Olga Popova 152r; Paul Fleet 141; pippeeContributor 124; rangizzz 137r; Rashevskyi Viacheslav 91r; Robcartorres 70; Roman Sigaev 173; Rost9 46l; sakkmesterke 189cl; science photo 112; Sebastian Kaulitzki 64l; Sebestyen Balint 93; Sergey Uryadnikov 63bc; Solodov Aleksei 90l; Steffen Foerster 66; Stock_Good 120; StockPhotosArt 65c; Syda Productions 103l; Sylvie Bouchard 90r; tbc 205l; Thomas Bethge 33r; Tony Baggett 70; Triff 83; twiggins 76; vchal 53c; vectorEps 87; Vilor 37c, 37r; Vlad_Nikon 156; Weerachai Khamfu 147; weknow 151; Willyam Bradberry 79; Yuliyan Velchev 171r; Yurchenko Yulia 57; Zeynur Babayev 51bg. **University of Chicago Medical Centre** 20. **University of Edinburgh Art Collections**. © the estate of Ruth Collett 34l. **US National Library of Medicine** 51. **Wellcome Collection** 7c, 24. **Wikimedia Commons** 4r, 6, 7l, 7r, 8l, 9c, 10, 11c, 12l, 12r,13r,16, 21, 26, 30, 31r, 32, 36, 38, 54, 55l, 63, 72, 75l, 75r, 80, 88, 90c, 102, 103c, 106, 109 inset, 122, 123, 125r, 126, 128, 130, 136, 137c, 138, 144, 145l, 150, 152, 154l, 154r, 159, 160, 163r, 164, 167c, 204; Acc. 90-105 - Science Service, Records, 1920s-1970s, Smithsonian Institution Archives 174, 182; American Institute of Physics, Emilio Segrè Visual Archives 185; Bundesarchiv, Bild 183-R57262/unknown (CC-BY-SA 3.0) 157r; Cush 178; Dinoguy2 (CC-BY-SA 1.0) 25r; Esculapio (CC-BY-SA 3.0) 8r; Harvard-Smithsonian Center for Astrophysics 183; Jaime A. Headden (CC-BY 3.0) 24r; NEON_ja (CC-BY-SA 2.5) 31l; Smithsonian Institute 202; T. Michael Keesey (CC-BY 2.0) 55r; TEM of D. radiodurans acquired in the laboratory of Michael Daly, Ur̲̲̲̲̲̲̲̲ 33 inset; The Catholic University of Leuven 189l; The Royal Society 3190106360690l s and Records Administration 175; Web of Stories (CC-BY-SA 3.0) 18.